The Comprehensive Guide to Wireless Technologies

By Lawrence Harte, Richard Dreher, Steven Kellogg, and Tom Schaffnit

Published by:
APDG Publishing

202 No. Main Street
Fuquay-Varina, NC 27526 USA
1-919-557-2260, 1-800-227-9681
Fax 1-919-557-2261
email: success@apdg-inc.com
web: www.apdg-inc.com

APDG

Library of congress: 97-078141

International Standard Book Number: 0-9650658-4-7

ACKNOWLEDGEMENTS

We thank all the gifted people who gave technical and emotional support for the creation of this book. Experts from manufacturers, service providers, trade associations, and other telecom related companies gave their precious personal time to help us and for this we sincerely thank and respect them.

These experts were: Mary Thurber and Craig Stanhope of Jupiter Data; Fred Dietrich and Tara Komar of GlobalStar; Naomi Yeransian of Ardis; Roman Kikta of Nokia; Tim Stanley and David Danaee of AT&T Wireless; Reed Fisher of Oki; Jeffrey Hines of BT Alex Brown; Simon Ford of London First; P.J. Louis of True Position; Elliot Hamilton of Strategis Group; Bradley Eubank of PCIA; Jim Fallen of GTE Wireless; Yaz Mochizuki and Atsushi Bando of Casio; Roger Deeringer of Lucent; Susan McQuaid and Rhett Grotzinger of Trident; Tyler Proctor of Zsigo Wireless; Pat Sturman, Josh Kiem, and David Kurt of Motorola; Erik Stasik and David Ethridge of Ericsson; Tom Steiner of Professional Technologies Systems; Elizabeth Tolson of LCC; Dawn McLain of Samsung; and Jim Mullen of Hughes Network Systems.

Special thanks must go to the following people for supporting and having faith in us: Tom Harte, James Harte, John O'Briant, Bob and Susan Rourke, Jaig Rourke, Carlene Sommer, David Dreifus, Patrick McKee, Robert Buttner, Laurence R. Gerson of Eastern Communications, Michael Zapata of TEC, Mike Cromie of Ericsson, Ted Ericsson of Ericsson, Herman Bates of RST, Konny Zsigo of Zsigo Wireless, Harry Young of Young Ideas, Carlton Peyton of Snap Track, and Linda Plano of Sarnoff.

We would also like to thank the professionals at APDG, Inc. for their assistance in the production of this book: Nancy J. Campbell, Lisa Gosselin, Erika C. McManus, Judith M. Rourke-O'Briant, and Michael H. Sommer.

About the Authors

Lawrence Harte is the president of APDG, a provider of expert information to the telecommunications market. Mr. Harte has over 19 years of experience in the electronics industry including company leadership, product management, development, marketing, design, and testing of telecommunications (cellular), radar, and microwave systems. He has been issued patents relating to cellular technology and authored seven books and over 75 articles on related subjects. Mr. Harte earned his Bachelors degree from University of the State of New York and an MBA at Wake Forest University. During the IS-54 TDMA cellular standard development, Mr. Harte served as an editor for the Telecommunications Industries Association (TIA) TR45.3, the digital cellular standards committee.

Richard Dreher, the Director of Advanced Technology for Evolving Systems, has over 16 years experience in digital communications, including Local Area Network design, SS7 engineering, cellular equipment product planning and technical sales, and Advanced Intelligent Network and Service Creation product marketing. His entrepreneurialism led to start-up experience with a PCS wireless carrier as their network systems technology director. Richard earned his BSEE from the University of Colorado, is a registered Professional Engineer (P.E.), a senior member and past executive officer of the Institute of Electrical and Electronics Engineers (IEEE), and a published author in trade journals, text books, and magazines.

Steven Kellogg has over fourteen years marketing experience directly within the wireless industry. His career began in 1983, where he led start up operations for several early wireless retail mobile electronics stores. He has collaborated on marketing programs with AirTouch Cellular, AT&T Wireless, GTE Mobilnet, 360 Communications, PageMart and others. Mr. Kellogg has well over 2,000 hours of experience in developing research, and is considered an expert in the strategic marketing development of cellular, PCS and paging services. He has consulted and developed start up operations and marketing blueprints for both post-pay as well as pre-pay wireless applications in many diverse markets throughout the country. He is a frequent guest speaker, providing wireless launch expertise at the NWRA WIRELESS 97 conference in Washington D.C., as well as the WIRELESS OPPORTUNITIES 97 conference in New York. Mr. Kellogg has co-authored strategic marketing analysis reports, including "Reselling Wireless", as well as a frequent contributor to industry trade magazines, including *Business Telecom* and *RCR* magazine.

Tom Schaffnit, President of CUE Data Corporation, is a professional engineer with extensive business experience in identifying, evaluating, and implementing new telecommunications systems and services. Over the past several years, he has been especially active in the area of wireless telecommunications. He is currently working on the rollout of wireless data services on the CUE FM subcarrier network. Tom has an undergraduate degree from Purdue University, and an MBA from the University of Manitoba. His work experience in telecommunications includes technical positions with the Manitoba Telephone System, Telecom Canada, MPR Teltech, and the Nordicity Group. Immediately prior to accepting his present position with CUE, he was a Senior Manager with the Deloitte and Touche Consulting Group's Telecommunications Centre of Excellence in Toronto.

I dedicate this book to my parents: Virginia and Lawrence M. Harte, my children: Lawrence William and Danielle Elizabeth, the rest of my loving family, and Tara.
- Lawrence

I dedicate this book to my wife Darcy and my children Aubrey and Austin as well as my extended Christian brothers and sisters who all support my ambitions with love and without judgement.
- Richard

I would like to dedicate this book to the millions of people who will have their lives enriched through the application of the new wireless services, such as those discussed in this book. As well, I would like to invite innovators everywhere to imagine, invent, and implement new services and new technologies which will require totally new books in the near future.
- Tom

CONTENTS

PREFACE

In 1998, there were over 680 million people using various types of wireless technologies. Wireless products and services that are being introduced to the market have advanced features and services, that enhance or replace alternative wired solutions. This book is a guide which provides the big picture of the major wireless technologies, history, market projections, basic operation, key industry terminology, and future trends.

This book offers an introduction to existing and soon to be released wireless technologies and services. It covers what's new in wireless from cellular to satellite. Each wireless technology has its own unique advantages and limitations which offer important economic and technical choices for managers, salespeople, technicians, and others involved with wireless telephones and systems. *The Comprehensive Guide to Wireless Technologies* provides the background for a solid understanding of the major wireless technologies, issues, and options available.

The book explains each technology and its services by using over 75 illustrations that have simple descriptions without formulas. Many of the industry buzzwords are defined and explained. These chapters are divided to cover specific technologies, services and trends and may be read either consecutively or individually.

Chapter 1. Introduction to wireless. Provides an introduction to basic wireless technology and industry terms. It covers who controls and regulates the wireless industry. A basic definition of each of the major wireless technologies and services is included. This chapter is an excellent basic introduction to wireless technology.

Chapter 2. Wireless applications. Describes the applications associated with wireless communications. This chapter gives a broad overview of the different types of wireless services and how they operate. This includes voice, data, remote control, point to point and broadcast services.

Chapter 3. Wireless basics. Explains the fundamentals of wireless technology and terminology. This includes how the radio frequency spectrum is divided, the basics of radio frequency transmission and modulation, antennas and radio networks.

Chapter 4. Land mobile radio. This chapter provides an overview of various specialized mobile radio (SMR) systems including integrated Digital Network (iDEN), EDACS, dispatch, and two-way radio. Significant features and trends in the SMR industry are explained.

Chapter 5. Cellular and PCS. Explains the different types analog and digital mobile telephone systems and their evolution. This chapter discusses the basic operations, attributes and services of cellular, wireless office and cordless systems.

Chapter 6. Wireless data. Covers the basics of packet and circuit switched data. Included is a description of various public and private systems and their messaging and data services.

Chapter 7. Paging systems. Descriptions of one-way and two-way paging systems and their history. This chapter also covers many new non-human applications for paging systems.

Chapter 8. Satellite communications. An explanation the different types of satellite communications is provided in this chapter. Included is public and private satellite systems including DBS, DSS, and very small aperture terminals (VSAT) Important characteristics of satellite systems are covered and future trends and applications such as high speed Internet access are discussed.

Chapter 9. Fixed wireless systems. Provides an overview of fixed wireless systems including wireless cable and wireless local area networks (WLAN). This chapter covers astounding market growth of wireless cable, the available and future technologies, and advanced services such as video on demand. This includes high speed licensed and unlicensed radio technology.

Chapter 10. Radio and Television. Explains the fundamentals of broadcast television and radio and future trends. High definition television (HDTV) basics are provided and future predictions are included. Find out about what new services may be available soon.

Chapter 11. Wireless Office and Cordless Systems. Describes the fundamentals of wireless office and cordless telephone systems. This includes market and basic technology information covering wireless private branch exchange (WPBX), public cordless, and other low power wireless telephone devices. Find out about how these technologies are evolving and what the recent changes and features have to offer.

Chapter 1

Introduction to Wireless

What is Radio

The ability to communicate via radio has changed dramatically over the past 5 years. Some people make money with these changes, some fear them and almost all people benefit from them. The physical properties of radio have not changed, but the applications have. Cellular phone systems offer paging, paging systems offer two-way voice messaging, wireless local loop systems offer phone and high speed data service and mobile satellite systems are bypassing the global telephone network. With all these recent changes, professionals and those involved in the telecommunications industry must continually become aware of competing applications of wireless technology. Until recently, to keep up to date with the latest systems, several books and courses were necessary to allow the industry player to understand the basics of each of the new technologies. This book is intended to provide the basics of wireless technology and provide an introduction to the latest in competing wireless technologies.

Radio comes from the word radiate and radiating electromagnetic energy is stored in our airwaves. Electromagnetic energy can be used to transfer information through the airspace by modifying or mixing the information with the electromagnetic radio waves. Wireless applications share one resource which all radio technologies share –airspace.

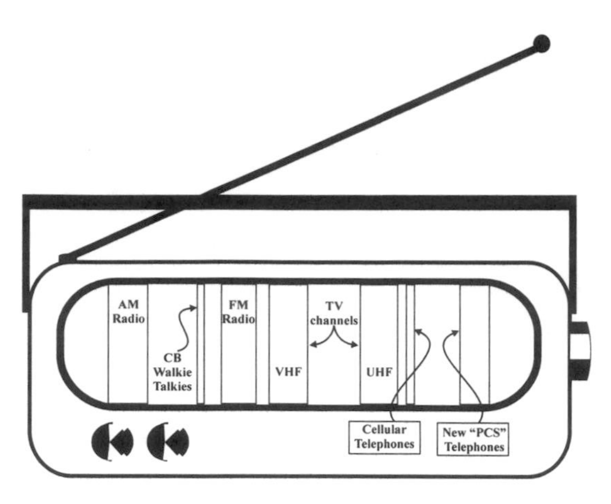

Figure 1.1 Radio Airwaves

There are hundreds of applications for wireless including AM/FM radio, broadcast or satellite television, pagers, baby monitors, garage door openers, military surveillance, aircraft guidance systems, police coordination, taxi dispatching and many others. Because radio signals operating on the same frequency interfere with each other, the use of radio signals must be managed like any other natural resource. The simple graphic in figure 1.1 can be viewed as an extended radio dial. It shows some of the wireless services and the relative frequency bands in which they operate.

To the left of the FM radio dial you have AM radio stations, Citizen Band (CB) "walkie talkie" radios and various types of unlicensed remote control devices (example: garage door openers). To the right, you have ultra high frequency (UHF) broadcast television (TV) channels (channels 21 through 69). The military has much of the white space between these public radio frequency bands. Cellular telephone frequencies come directly after the UHF television channel 69. The frequencies used for cellular telephone service were previously assigned to UHF television channels 70 through 83. At a frequency almost double normal cellular frequency, personal communications service (PCS) radio channels were created in 1994 to allow services that are similar to cellular telephone service to enhance competition. Notice how the newer applications move up the frequency scale as more sophisticated skills, to harness the "higher frequency" airwaves, are developed. Above the PCS radio channel frequencies come microwave radio services including digital satellite services (DSS).

These invisible airwaves are like the little waves-of-water which emerge in circular formation after a source of energy (a pebble) forces the water to respond to its presence. These little-wave-rings will fade away if another pebble is not dropped into the water. However, if you frequently drop pebbles into the water, the waves will continue. And if the frequency at which you drop the pebbles into the water is spaced equally in time, the wave pattern in the water will match that source frequency (the rate at which you drop pebbles). Like the water surface of a busy lake – with its collision of waves upon waves, electromagnetic waves ride on top of each other.

Fortunately technology has enabled people to “see” these electromagnetic waves. The waves are labeled by their frequency of movement, in waves per second. Since they were first generated by the German physicist Heinrich Hertz in 1888, we call a cycle (wave) per second a Hertz (Hz). Figure 1.2 shows how to measure a wave in Hertz. In this diagram, there are three cycles in 1 second which equals a frequency of 3 Hertz. Radio waves typically have several million cycles per second which is called a Megahertz (MHz).

All radio transmission devices must be operated in their assigned range of frequencies to avoid collision with other radio transmitters operating in their designated frequency range. For example, FM radio stations only transmit frequencies which fall in the boundaries of an FM radio dial. Likewise FM radio receivers “listen” to the airwaves and use what is known as a band-pass-filter, to filter out unwanted frequencies. The filter then passes through a ***band*** of frequencies allo-

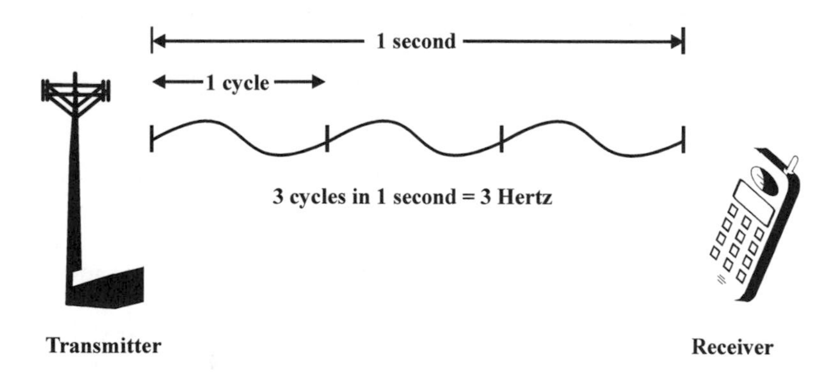

Figure 1.2 Measuring a Wave in Hertz (Hz)

cated to FM radio. Further tuning of the FM radio (a second stage band-pass-filter) will filter down to a single frequency; like a favorite radio station.

Because all applications share the same airspace, the radio spectrum has been divided into measurable frequency band ranges managed by the governments of the world. If any source radiates electromagnetic energy, they are subject to the rules of the Federal Communications Commission (FCC) in the US and department of communications (DOC) in other countries. Even unlicensed radio devices such as garage door openers are subject to government regulation.

Regulation of Wireless

In 1912, Congress passed the first airwave licensing law and nearly a thousand licenses were issued, many to colleges and universities. As a flood of license applications continued, the secretary of commerce established the Federal Radio Commission (FRC) in 1927 to settle the airwave chaos, including interference due to licensed and unlicensed (unknown) radio operators broadcasting at overlapping frequencies. Today people don't think twice using wireless devices, like the latest invisible fencing, portable telephones, or remote "keyless" entry systems for cars. Yet with all these new wireless *applications*, the frequency spectrum available to the general public is running out.

In the United States, responsibility for allocating the radiowave portion of the electromagnetic spectrum was transferred in 1934 from the FRC to the newly created Federal Communications Commission (FCC). Since then the FCC has overseen division of the airwaves in a complicated political process for the benefit of the general public. The FCC will over the next 15-20 years reallocate spectrum for public use as required by the Omnibus Reconciliation Act of 1993. The majority of this reallocation will be taken from the spectrum currently allocated to the federal government and defense department.

The FCC (US) or DOC (rest of world) not only divides up the frequency, they also regulate what the frequencies must be used for. For example, the frequencies allocated for FM radio must be used for a combination of music or news and public information. The radio station cannot broadcast a secret "Morse-code" to a following of undercover militia! Neither can a "paging service" use one or all of their frequency channels to broadcast stereo music. This is enforced by the "licensing" of the airwaves. If a licensed radio service provider ("carrier") of a particular band of frequencies violates their license requirements, their right-to-use the frequency will be taken away from them.

Licensed

Clearly the largest amount of radio spectrum is designated for particular purposes. As can be seen in Chapter 3, the frequency spectrum is divided by "type" of usage, the licensing of these frequency bands reflects this purpose. Originally the sole purpose of licensing was to avoid collision in the airwaves; however, the FCC has recently decided to cash in on this precious resource by establishing a license fee based on an auction process.

Earlier awardees of cellular licenses, for example, were only charged a minimal documentation fee by the FCC, as the awardee does not own the frequency(s), only the right-to-use it under the license guidelines. The awardee would build a cellular system per the license requirement and then after a couple months or years, they would sell their right-to-use "their" license and radio system for millions of dollars to the highest paying company. What the FCC licensed for a small documentation fee, was resold to make the lucky awardee a millionaire.

To the exception, some radio frequency bands are designated for general public use, or usage of a non-commercial basis such as the frequencies allocated for Amateur Radio and Citizens Band (CB) radio. Initially, both of these public frequency bands required a license to operate. Now, certain users of amateur radio and CB-Radios no longer require a license to operate.

Unlicensed

While there are many radio transmitter and receiver devices that do not require a license to operate (called "unlicensed"), all wireless device usage is governed by the FCC. General purpose usage of transmitters such as remote control devices that do not require a license to operate (such as garage door openers or radio controlled toys) are regulated by specifying to the manufacturer the frequency, maximum radio transmitter power and method used to control the transmission of the radio signal. When these devices are manufactured to the FCC requirements, the amount of interference they cause to nearby devices is reduced and essentially controlled.

The unlicensed frequencies are designated under Part 15, Radio Frequency Devices, of the FCC regulations. Part 15 basically states that you must tolerate any interference from any other device operating in the same frequency. In other words, a person cannot ask for an impending transmitter user to "turn off" or change their transmitter frequency. Some devices such as cordless telephones and wireless LAN's automatically avoid interference by listening for other radio signals prior to transmission. If another radio transmission is detected, the cordless device automatically changes its channel frequency; if not, an unwanted conversation may be heard on the same channel. This is termed "frequency hopping" in most technologies that use the technique.

Recently the FCC allocated additional unlicensed airwaves designated for personal communications services (PCS). To avoid airwave collision here, the FCC has named a neutral non-profit consortium of leading telecommunications and data equipment manufacturers, as coordinator for the unlicensed spectrum. This PCS Unlicensed spectrum is being used by computer companies for point-to-point communications links, in a local area network (LAN) environment as an example.

Categorizing Wireless Services

The main way to categorize wireless services is by the types of services that are offered by the carriers of the radio waves. Radio waves are used to carry information for various types of services. These services can be divided into broadcast, two-way radio and point-to-point services. Broadcast services (such as television) allow many receivers within a radio coverage signal to listen or watch the same information carried by a radio signal. Two-way services allow a defined group of receivers (such as taxis in a city) to send and receive voice communications or messages with independent operation. Point-to-point services allow controlled (private) communications between two fixed or portable locations (such as a cellular telephone call).

Broadcast Radio

Broadcast radio involves the one way transmission of the same information to all receivers in a geographic area. Broadcasters usually do not know the recipient of information. In this fashion the content of the information is generally public, where the cost of doing business is recouped by advertisers who broadcast their goods or services over the air. Radio and television programming has played a dominant role in information and entertainment delivery. These services have been provided using technology that was developed decades ago. The transmission standard is still analog. Analog is still proficient at distributing signals such as voice and video, however it does not allow digital data to be transmitted easily.

In today's information age, the requirement for broadcast data has the industry currently poised to deliver high quality voice, video and data services using conventional broadcast frequency allocations. These include the development of high-definition television (HDTV), digital audio broadcasting (DAB), and FM high-speed subcarrier (FM HSS) data services.

Figure 1.3 shows how broadcast radio works. In this diagram, one high power transmitter is located at a high point above the city. The transmitter may be broadcasting at over 50,000 watts to allow a radio signal to reach many homes and businesses within a 50 mile radius. Ordinarily, any receiver that is capable of tuning into the broadcast signal can receive and listen to it. However, some systems (such as movie channels) are scrambled so only receivers with a signal decoding capability can listen or view the signal.

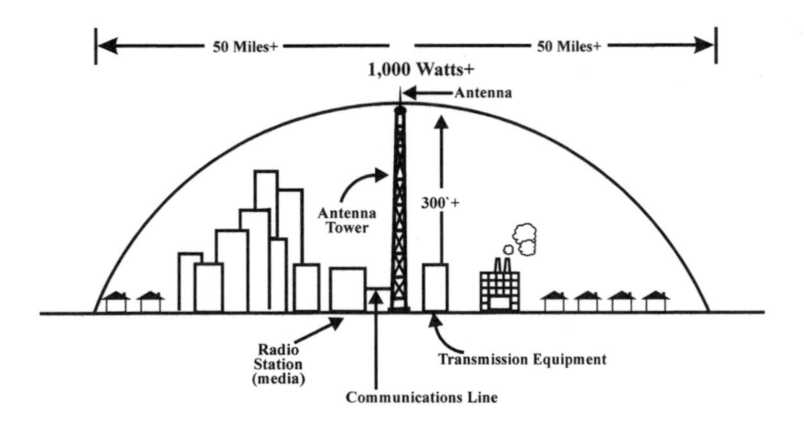

Figure 1.3 Broadcast Radio Service

Two-way Radio Communications

Two-way radio communications consist of a wide variety of mobile radio systems, ranging from a simple pair of handheld citizen band (CB) "walkie-talkies" to elaborate wireless backbone radio repeater systems. Two-way radio communications also include non-terrestrial services such as: aeronautical mobile radio (communications with air-craft), maritime mobile radio (communications with ships at sea), and the new mobile satellite service (mobile communications via satel-lites).

Two-way radio communications systems are commonly referred to as "Land Mobile Radio" (LMR) systems. Land mobile radio can be divid-ed into private and public systems. Private systems are owned by the operator who normally communicates with a small group of users. These users may be building maintenance workers or local delivery services. Public land mobile radio systems are owned by a company that offers services to many users that belong to different groups. These groups include taxi and trucking companies. Public wireless telephone systems, licensed by the FCC, are ruled under the new Commercial Mobile Radio Service (CMRS). These two-way mobile communication applications fill a general need for mobile and person-al communications that has made a significant impact on business productivity, personal safety, and general social behaviors.

Figure 1.4 shows a basic private two-way land mobile radio system. In this diagram, a dispatcher (maintenance coordinator) uses a moderate power radio transmitter that can send a radio signal to several mobile radios operating within the building. The transmitter regularly uses

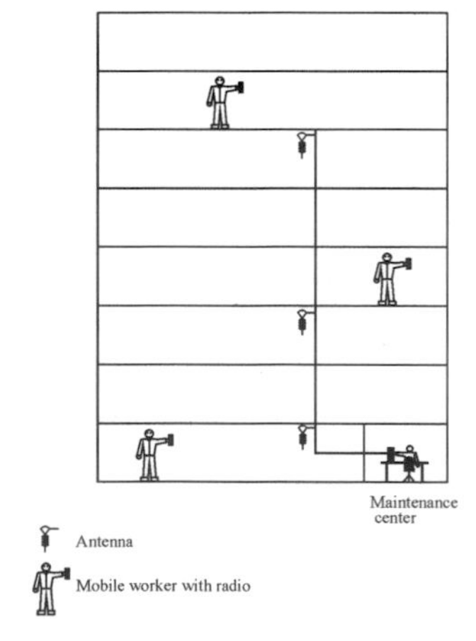

Figure 1.4 Private Two-Way Radio System

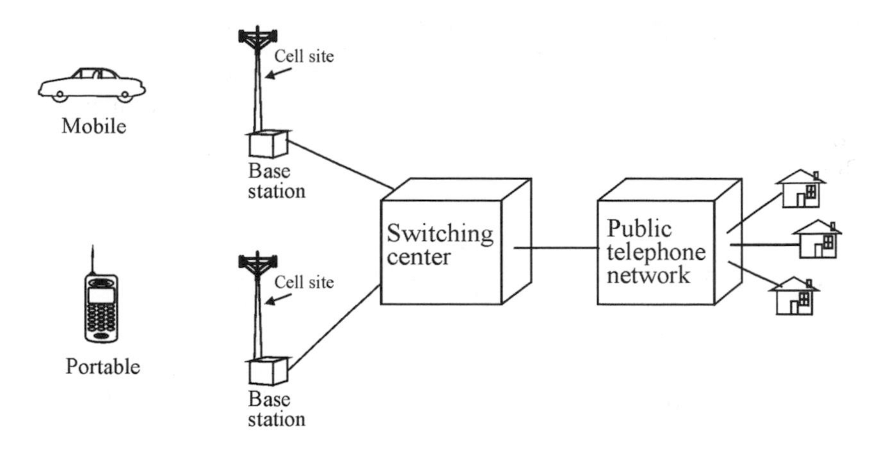

Figure 1.5 Public Cellular Radio System

less than 10 watts of power which limits the radio signal to reach only a few miles. As a rule, portable radios are capable of tuning into one or more frequency and may communicate with the base or each other. In some systems, low power mobile radio signals are received by the base and retransmitted by the base (or other repeater) so all the portable radios can receive the message.

In public cellular systems, many mobile telephones communicate through nearby radio base stations and their central switching system which then connects to other wired and wireless telephones. Mobile telephones can be portable (handheld), transportable (phones in a bag or case) or mobile (mounted in a vehicle). They can change frequency and power level and are usually controlled by messages received from the cellular system. Mobile telephones communicate through radio base stations that are located within a few miles. Each base station commonly transmits and receives on several fixed frequencies. The base station converts and transfers the radio signals between the mobile telephone and switching system via dedicated wires or a microwave link. The switching system connects the calls to the public telephone network or to other mobile phones operating in the system. Figure 1.5 shows a basic public two-way cellular mobile radio system.

Point-to-Point Services

One categorization of point-to-point communications involve the transmission of signals from one specific point to another, as distin-

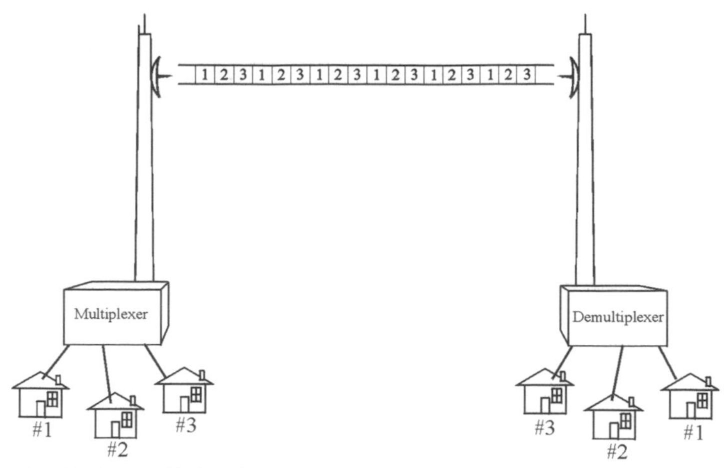

Figure 1.6 Point-to-Point Services

guished from broadcast transmission which blankets the general public. Microwave radio relay links are a common type of point-to-point communication.

Point-to-point service is sometimes categorized as the end-to-end connectivity for sending information. This categorization is not technology specific, nor regulatory specific, it relates to applications where information is deliberate and routed between two specified destinations. Sometimes short message service (SMS) that is sent between specific users is called point-to-point messaging.

Figure 1.6 shows point-to-point wireless service. In this diagram, multiple communications channels are multiplexed (time shared) to create a combined information signal. This combined signal modulates a radio carrier that is sent to a directional antenna. This antenna sends a radio signal in a focused beam to another directional antenna which receives the signal. The received signal is demodulated and the multiple channels are separated (de-multiplexed) back into their original parts.

Wireless Applications

Voice

Voice communications allows a person or audio device to communicate to one or more radio receivers by transmitting a radio modulated sig-

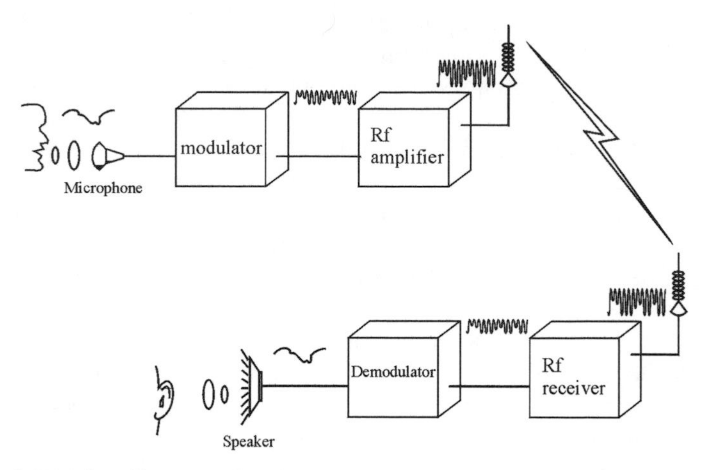

Figure 1.7 Voice Communication

nal. Voice communications typically involve real time communications where both the sender and receiver must be communicating at the same time.

Figure 1.7 shows a typical voice service application where a mobile radio is communicating with another mobile radio. As the user talks, their audio signal (sound pressure) is converted to electrical signals. These audio electrical signals (called the baseband) transfer the information by modifying (modulating) the radio transmitter frequency (called the broadband) signal. This low level radio signal is boosted by an RF amplifier in the transmitter section and converted to electromagnetic waves by the antenna. A nearby mobile radio receives the radio signal by using a frequency filter and compares the modified incoming radio signal to an unchanged frequency to extract the original audio electrical signal (called demodulation). The audio signal is then converted to the original acoustic sound by a speaker.

Data

Data communications involve the transfer of digital information between a computer or other digital device to one or more radio receivers by transmitting a radio modulated signal. While it is necessary to have a wireless data transmitter and receiving communications device to operate at the same time to transfer the data, the user of data communications (the computers) ordinarily do not require real time end to end connectivity. For example, if during a file transfer the computer is asked to wait for 5 to 10 seconds for new data, this may be acceptable for data transfer and unacceptable for voice communications. This allows wireless data communications to use retransmit request messages to ensure messages were received correctly.

Digital information regularly has only 2 levels (on and off). On signals are usually called a 1 and off signals are usually called a 0. The transfer of digital information by radio waves involves conversion from digital signals to analog radio waves. All radio waves can experience distortion. Unlike voice communications where the users can normally tolerate small amounts of distortion (noise) of the audio signal, digital information cannot as a rule tolerate distortion. Distortion could result in the misunderstanding of information or the incorrect reception of 1's and 0's. To ensure digital information is correct and complete, various error detection and error protection schemes are used. These schemes send add data bits, which increases the overall message length and reduces the maximum data transfer rate. This allows the systems to determine if the information is complete and correct. If the systems determine that errors have been received, a re-transmission request can be requested. These techniques are not unique to wireless data transmission and have been in use for many years by the information industry. (For the protection of digital information, additional digital bits are used which reduces the maximum data transfer rate of the system and retransmission requests may be sent when data is determined to be distorted.)

Digital information can be transferred via two basic methods; continuous and burst. Continuous data transmission is called "circuit switched" and burst transmission is called "packet switched." For continuous data transmission (such as a large data file), the address is sent first and a continuous path is dedicated for data transmission. This path is maintained even if the source of data information is temporarily interrupted. For packet data transmission (such as a credit card validation request), the data information (the message) is divided up into small packets of digital information (such as 100 characters at a time) and each packet is given a sequence number and a destination address. These packets migrate through an interconnected network independently and are reassembled into the original message when they reach their destination.

Figure 1.8 shows a typical data service. In this example, a computer is sending a file. As the computer sends digital information (1's and 0's), their baseband electrical signal (high or low voltage) transfers the information by modifying (modulating) the radio transmitter frequency (broadband) signal. This low level radio signal is boosted by an RF amplifier and converted to electromagnetic wave by the antenna.

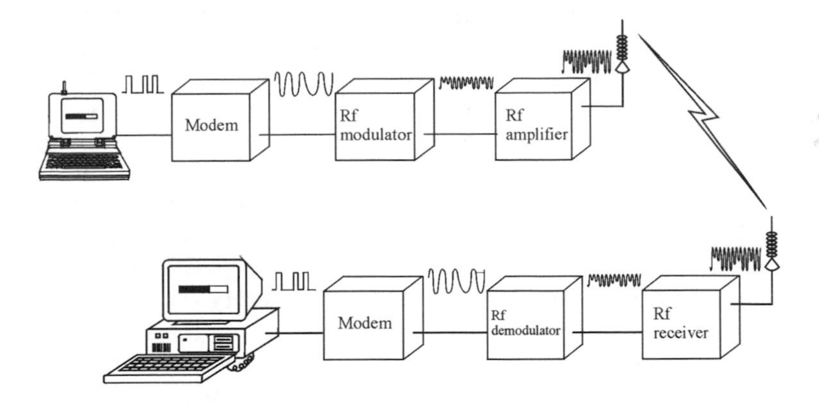

Figure 1.8 Wireless Data Communication

Another wireless data device receives the radio signal and many others from its antenna. Its receiver selects the correct radio signal by using a frequency filter and compares the modified incoming radio signal to an unchanged frequency to remove the audio electrical signal (called demodulation). The audio electrical signal is converted to the original digital by a modem (modulate - demodulate) and routed to another data device.

Video

Video communications allow a broadcasting company to send a visual signal (constantly combined visual and audio) to many radio receivers (typically televisions) by transmitting a high power radio signal that is modulated with visual information. Video communications commonly are sent in one direction and delays and distortion can be tolerated.

Video signals are very complex (rapidly changing). Because of this complex signal, the required amount of frequency bandwidth is much larger than typical audio (cellular and radio). For example, 200 standard cellular radio channels (30,000 Hz wide) can fit into the radio bandwidth allocated to 1 television channel (6,000,000 Hz each).

Figure 1.9 shows a typical video service. In this example, a television camera converts an image and audio sounds to electrical signals. The video signal is created by a camera scanning the viewing area line by line. At the beginning of each line scan, the camera creates a synchronization pulse and the image (light level) is created by varying the electrical signal level after the synchronization pulse. The audio signal is created by using a microphone. These video and audio electrical signals are combined to form a composite video electrical signal. The composite video signal (baseband) modulates the radio transmitter frequency (broadband) signal. This low level radio signal is amplified to a very high power level for transmission. A video receiver (ordinarily a television) receives the radio signal and many others from its antenna. It's receiver selects the correct radio signal by using a variable frequency filter (television channel selector) that demodulates the incoming radio signal to create the original video and audio electrical signals. The video signal is connected to a display device (usually a picture tube) and the audio signal is connected to the speaker.

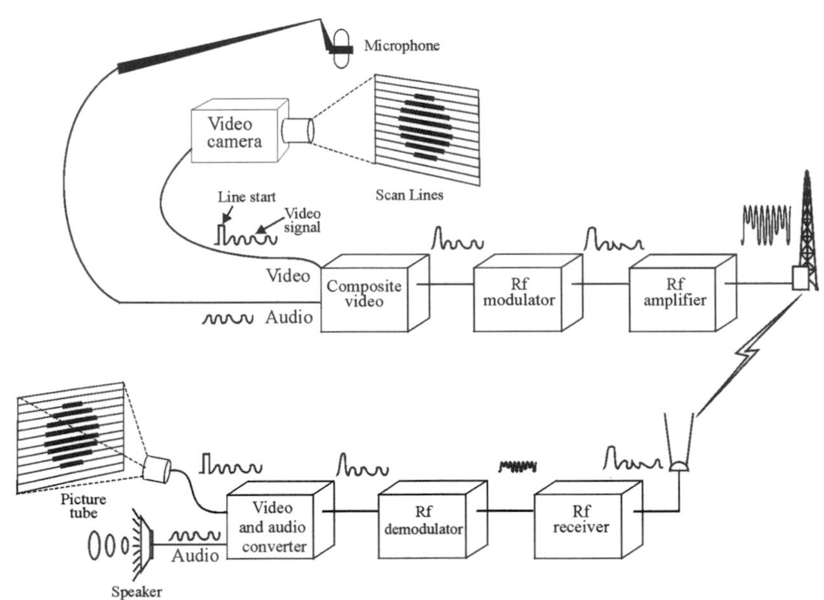

Figure 1.9 Video Communication

Chapter 2

Wireless Applications

Alexander Graham Bell predicted that someday every city in the country would have a telephone

Wireless services such as mobile telephony, radio broadcast and television are commonplace throughout the world. The types of services vary dependent on the application. Wireless applications are a combination of the physical transmission of radio signals and rules or programs that coordinate the operation of the communication system. Wireless applications are created to produce benefits for their users. Successful applications include mobile telephones, land mobile radio, paging, wireless data, fixed wireless, remote control, point-to-point communications, broadcast radio and television.

Mobile Telephones

Mobile telephones connect people to the public switched telephone system (PSTN) or to other mobile telephones. Mobile telephone service includes cellular, PCS, specialized and enhanced mobile radio, air-to-ground, marine, and railroad telephone services.

The first mobile telephone system in the United States began in St. Louis, Missouri in 1946. By 1947, more than 25 cities in the United States had mobile telephone service available. The systems used a single high-power transmitter for the base station *in the center of a metropolitan area*. Coverage was provided for 50 miles or more from the transmitter. These initial systems used a human operator at the base station to manually connect the mobile user with the landline network. In most of these systems service was very poor because too many customers (called subscribers) shared each radio channel (called loading). It was not uncommon to have busy channels over 50% of the time. Despite this poor service, it revolutionized the definition of telephone service and priority was given to police and ambulance service. The waiting list for mobile phones in some cities was more than 7 years. This type of system was improved many times and the last upgrade, called improved mobile telephone service (IMTS), was introduced in the mid 1960's. While there may still be some original systems in operation throughout the United States, new equipment for these systems is not currently being produced. It has been replaced with cellular systems.

Cellular and PCS

Cellular and PCS mobile telephone systems allow mobile telephones to communicate with each other or to the public telephone system through an interconnected network of radio towers. This differs from the first mobile telephony systems which utilized a single high-power transmitter with a tall transmit antenna that could communicate to high power mobile phones up to 40 or 50 miles away. "Cellular or PCS" was born when the traditional single antennae coverage area was divided into a group of smaller radio coverage areas (called "cells"). The same large geographic radio coverage area (typically over 5,000 square miles) used for IMTS mobile telephones could be covered by many small radio coverage cells (each covering from 5 to 500 square miles). Within each cell, a low-power transmitter was used to provide coverage only within that cell. When a mobile phone moved from one cell into the other, the radio signal would be transferred to the next cell radio coverage area (called "handoff"). The transfer is coordinated by an automated switching system.

The cellular connection refers only to the radio frequency (RF) portion of the wireless system. A complete network is required to maintain a connection to the PSTN and other mobile telephones that are operating in the system. The PSTN is also known from a regulatory standpoint as a common carrier telephone system that provides telephone service to the general public. The cellular connection involves a centralized telephone switch, commonly called a mobile switching center (MSC) linking the cells to each other, and to the PSTN.

Cellular systems have grown beyond most early expectations. In October of 1983, analysts at AT&T, which at that time was the largest company in the world, forecasted there would only be a total of one million cellular subscribers in service by the year 2000. By 1993, the end of the first decade of availability of cellular telephones, there were already 16 million cellular telephones in use. In 1998, there were over 54 million with an approximately 28,000 new users coming on line per day!

Definition of PCS

In the late 1980's, cellular service providers (called "cellular carriers") determined that the rapid growth rate of the cellular telephone marketplace would quickly exceed their ability to provide service to all the new customers. Either the technology, business structure or regulations would have to change so service could be available to all who applied for service. The key limiting factors in the equation of growth were the amount of radio frequency that was available to cellular service providers and the efficiency of the analog cellular technology that was deployed in 1983. The industry first asked the FCC to reallocate more frequencies for cellular. The FCC was told the demand for more cellular was based on the unprecedented acceptance of "cellular" usage for personal use, rather than for business use. This adoption of cellular phones for personal communications was in part due to the decreased size of the phones and the relative cost of the service. With such rapid proliferation in personal usage of cellular, a new market opportunity was realized – to provide the most competitive, value added services, aimed at personal communications. The FCC responded in 1986 with an additional 10 MHz allocation of frequency bandwidth for the cellular carriers. Even with the additional frequency bandwidth allocation, cellular carriers continued to experience difficulty meeting the high market demand for cellular telephones. In 1989, work began on developing a new efficient technology that could meet the market demands for mobile telephone service and advanced features. There were several analog and digital technologies proposed and some of these have reached the commercial marketplace.

While the initial analog systems were being upgraded to digital to support the new growth, the FCC began to consider reallocating new frequencies to support the new growth. Unlike the previous bandwidth that was given to the existing cellular carriers, these new frequencies were designated to be licensed for "Personal Communications Services (PCS). PCS is a term originally used by the FCC to identify the new radio frequencies at around 1.9 billion Hertz (called Giga-Hertz) that would be used for cellular-like service. Since then however, there has been a redefinition of the term PCS.

The industry began redefining PCS as something new, something well beyond the allocation of invisible airwaves. The definition of PCS turned into a battle of defining the "new paradigm" of the way people communicate. Definitions include touting how PCS is about communicating with people rather than focusing on the technology. After all the new owners of the PCS licenses paid big money to the FCC for the right-to-use these PCS frequencies, they had to convince their shareholders that their company is the company that will attract all the new wireless users. In a nutshell, marketing got a hold of the new term and publicized the benefits of the all new digital PCS systems and not the frequencies they bought.

While analog networks were being upgraded to digital to support the new growth, the FCC was reallocating new frequencies for PCS service. While it is true the new PCS frequencies are utilizing digital technologies, the old cellular networks were already in the process of upgrading to digital. The old cellular networks gone digital, now posses the same capabilities of the new systems using the PCS frequencies.

In the early market introduction phase of the new PCS licensed operators, the number of installed radio towers was much less than cellular service. This resulted in available radio coverage becoming the biggest obstacle. To increase the total wireless coverage area offered by the new PCS licensed operators, dual frequency phones (called dual band phones) are being used. These carriers may use dual band phones to provide their customers access to both cellular and PCS frequencies for better "wireless" service. Dual band phones also allow an established cellular carrier to acquire new PCS frequencies to be used for additional radio channel capacity in the same market area. Figure 2.1 shows a dual band phone that operates at both the existing cellular and new PCS frequencies.

Figure 2.1 Dual Band Cellular and PCS Telephone
Source: Ericsson

Aircraft Telephones

Aircraft telephones allow people on an airplane to initiate telephone calls to the public telephone system through connection via land based or satellite systems. Recently, some aircraft telephone systems have been upgraded to allow calls to be received on the airplane.

Aircraft telephone systems are ordinarily a hybrid wireless system that is a terrestrial wireless system (land based) combined with satellite service. The terrestrial system is used to connect telephone calls when the aircraft is above land and is within distance of a ground transmitter. For the terrestrial based system, the phone handset in the airplane is connected to a transmitter in the plane's belly that connects the call down to one of the ground antennas located strategically throughout the country. The call is routed to a ground switching station that connects the call to the receiving party.

The satellite system is used mainly over the water, where calls are out of reach of the ground antennas. For the satellite-based system, the phone handset on the plane is connected to an antenna on the top of the plane that connects the signal up to an orbiting satellite. The call is then sent down to earth by the satellite frequencies to its satellite earth station, then to one of the main ground switching stations that routes the call to the PSTN.

Aircraft phone systems normally have handsets in a common area or handsets that are located in the back of passenger seats. If the hand-

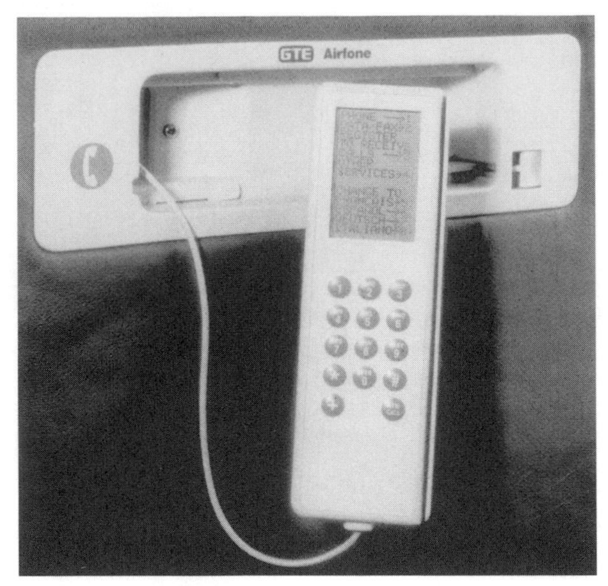

Figure 2.2 Public Aircraft Telephone
Source: GTE

set is located in the seat, some aircraft phone systems allow incoming calls. Is "hello, seat 8E" the proper way to answer the phone in front of you? For example, for someone to reach you on a GTE Airfone system, you first get an "Airfone Activation Number" by registering on the phone near you. The person placing the call from the ground dials 1-800-AIRFONE and follows the voice prompts to enter your number. Figure 2.2 shows a public aircraft telephone.

Mobile Satellite

Mobile satellite telephone service allows customers to use specialized satellite mobile telephones to communicate in any part of the world to the PSTN through the use of communication satellites. Commercial communication satellite services began in the mid-1960's with the establishment of Intelsat, a multinational organization with well over 130 member nations today. An organization known as the Communications Satellite Corporation (COMSAT) also was established in the early-1960's and became the United States' representative in Intelsat. These first commercial applications of satellites provided international telephone and television program transmission, primarily between the United States and Europe.

The satellite systems used for portable phone service via satellite are known as communication satellites. Using communication satellites have two advantages over traditional cellular technologies. One is the ability for a single satellite to broadcast the same signal covering up to one third of the Earth's surface. The other is the ability to provide reliable communication between any pair of points independent of a local telecommunications system.

There are three basic types of satellite systems; geosynchronous earth orbit (GEO), medium earth orbit (MEO) and low earth orbit (LEO). GEO satellites hover at approximately 22,300 miles above the surface of the earth. GEO satellites revolve along with the earth once a day, they appear stationary with respect to the earth. The high gain antennas used to receive signals from 22 thousand miles away (usually called "dish" antennas) are pointed directly toward the satellite. MEO satellites are located closer to the earth than GEO satellites and do not as a rule require high gain antennas. This is important as MEO satellites revolve around the earth several times per day and fixed antennas cannot be used. The newest satellite technology being deployed is LEO satellites. LEO satellites are located approximately

Figure 2.3 LEO Mobile Satellite Telephone
Source: Globalstar

450 miles above the surface of the earth. Because these satellites are relatively close to the earth, portable phones with smaller antennas can be used. Figure 2.3 shows a LEO mobile satellite telephone.

Wireless Office

Wireless office telephone systems are used in a business environment to provide similar features as a private branch exchange (PBX) with the ability of mobility throughout the office area. The wireless office commonly begins with a specialized wireless private branch exchange (WPBX) that has been adapted for wireless. While more complex than a home cordless telephone, it is not typically as complex as a complete cellular telephone system.

The WPBX telephone radio coverage area is usually within one or more company buildings or on a campus. The more popular WPBX systems use unlicensed frequencies with a protocol available only to the manufacturer of the WPBX. Ordinarily, WPBX telephones cannot be used outside the established campus.

These private WPBX systems use small wall mounted antennas, and like cellular, the space is divided to provide adequate capacity for the expected usage. WPBX telephones, like the one shown in Figure 2.4, have become commonplace in many hospitals and warehouse environments where the staff is primarily walking around to do their job.

Figure 2.4 Wireless Office Telephone
Source: Spectralink

Recent hybrids have been developed whereby the telephone handset has two technologies built into the operation of the phone. When the telephone is inside the WPBX coverage area (preferred) it acts as a private phone; when outside the WPBX coverage area, the phone has the ability to send and receive calls on the public cellular system, incurring airtime charges as any other cellular user.

Cordless

Cordless telephones are low power wireless telephones that connect calls to a customer's standard telephone line through the use of a home base station. The home base station communicates with the cordless phone and coverts the radio signal to the audio signals that are sent and received from the standard telephone line. Cordless telephones have evolved from simple radio extensions that operate on a single radio channel to multi-channel advanced digital wireless telephones.

Most home cordless telephones conform to the FCC's designated Part 15 unlicensed radio frequency regulations. Because so many homes operate cordless phones, each manufacturer must build-in circuitry to minimize the interference caused by other Part 15 devices. The original cordless phones use a very crowded Part 15 frequency band (around 49 MHz) utilizing analog radio wave modulation. Recently the FCC allocated new Part 15 unlicensed frequencies for cordless phones in the 900 million Hertz (Mega-Hertz) frequency range.

Many cordless phones that operate in the new "900 MHz" frequency band use the 900 MHz name as a part of their market differentiating

Figure 2.5 900 MHz Cordless Telephone
Source: Sony

features. Similar to the older cordless phones, 900 MHz cordless phones may use analog or digital frequency modulation. Digital cordless phones are normally better than analog cordless phones, and yet a 900 MHz analog might outperform an original 49 MHz digital cordless phone. Figure 2.5 shows a 900 MHz digital cordless telephone.

Land Mobile Radio

Land mobile radio (LMR) consists of a wide variety of mobile radio systems, ranging from a simple pair of handheld "walkie-talkies" to digital cellular-like systems. Land Mobile Radio (LMR) includes radio service between mobile units or between mobile units and a base station. LMR is licensed by the Federal Communications Commission (FCC) under several categories, including: private radio, Part 95 Personal Radio (Citizen Band (CB); where the owner operates within a set of 40 frequencies near 27 MHz, talking with other CB users), and the new Commercial Mobile Radio Service (CMRS); where the user owns his radios, but subscribes to repeater and other services provided by a system operator. Figure 2.6 shows a typical two-way land mobile radio.

Automated land mobile radio systems are divided into two categories; specialized mobile radio (SMR) or Enhanced SMR (ESMR).

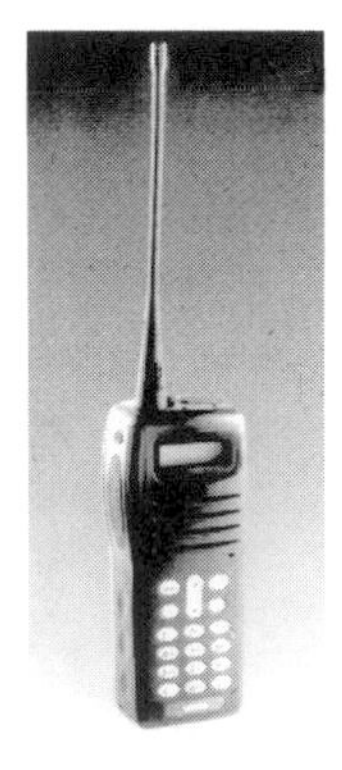

Figure 2.6 Two-Way Land Mobile Radio
Source: Uniden

Figure 2.7 Dispatch Radio
Source: Motorola

Specialized Mobile Radio (SMR)

Specialized mobile radio (SMR) systems are used by taxicab companies, conventioneers, police and fire departments, places where general dispatching for service is a normal course of business communications. SMR radios, as shown in figure 2.7, are regularly designed to be rugged to survive the harsh environment. SMR radios can usually be programmed with a unique code. This code may be a individual code or group code (e.g. pre-designated group of users such as a fire department). This allows all the radios belonging to a group, or a subgroup of radios can be "paged" by any party in the group. A push-to-talk method is used during the dispatch call (page) or reply. This push-to-talk radio-to-radio communication efficiently utilizes the airwaves because of the bursty (very short transmission time) nature of the information. Figure 2.7 shows a typical dispatch radio.

The SMR operator may also offer what is called a trunked repeater system, which can also provide PSTN connections. Depending on the particular SMR equipment capabilities, the telephone interconnect may operate only in the half-duplex mode, which requires a push-to-talk operation by the SMR user.

Enhanced SMR (ESMR)

ESMR is a name applied to a class of SMRs that use advanced digital technologies to provide enhanced telephone-like services to the user. ESMR systems provide a wide range of telephone, dispatch, alphanumeric paging, fax and data services. From a consumer point of view, if

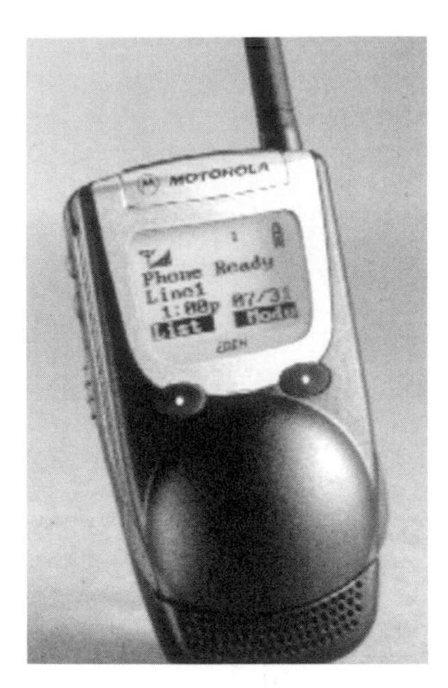

Figure 2.8 ESMR Radio
Source: Motorola

an ESMR service (such as NexTel Communications) is sold without the push-to-talk feature, the service would be indistinguishable from a traditional cellular service. The all-digital enhanced SMR system is a proprietary standard by Motorola called iDEN for integrated Digital Enhanced Network. ESMR networks are designed using large numbers of radio antenna sites and today compete head to head with traditional cellular carriers.

Traditional cellular is not included in the FCC's Land Mobile Radio classification. Cellular telephone service is considered (from a regulatory standpoint) an extension of the common carrier telephone system. Figure 2.8 shows a typical ESMR radio.

Paging

Paging is a method of delivering a message, via a public or private communications system or radio signal, to a person whose exact whereabouts are unknown. Users as a rule carry a small paging receiver that displays a numeric or alphanumeric message displayed on an electronic readout or it could be sent and received as a voice message or other data.

Paging began in 1949 with the allocation of frequencies exclusively dedicated to one-way signaling services. Subscribers used AM receivers, listened for an operator to announce their number, and then called the service to receive their messages. Selective addressing (the ability to choose one individual pager from the group) was introduced in the mid 1950's and FM was first used in an experimental paging system in 1960. Pagers with alphanumeric displays made their debut in the early 1990's. In addition to complete messages that can be sent and stored in these pagers, a number of other services such as stock market and sports score reporting have been developed.

Strong growth will continue, since the penetration rate (the number of customers compared to the population) in the United States at the end of 1996 was only about 16 percent. Usage is expected to increase to 56.2 million by the year 2000 and 92.2 million by 2005. One driving force for this expansion is the newest narrowband PCS frequency spectrum recently auctioned by the FCC. The new PCS spectrum allows for two-way paging. Two way paging will change the paging industry like microwave ovens changed the cooking industry.

One-Way Paging

A paging technology whereby the signal is sent from the base station to the paging unit only, without a return verification signal or other 2-way capabilities. Figure 2.9 shows a typical 1-way pager.

Figure 2.9 One-Way Pager
Source: Maxon

Figure 2.10 Two-Way Pager
Source: Motorola

Two-Way Paging

Two-way paging service refers to the paging system that has the ability to send paging messages and receive confirmation or responses from the portable paging device.

Two-way paging will effect change because of the host of potential interactive sessions capable between the pager and information providers. Whether it is as simple as confirming the loaf of bread is on the way, or as complex as a stock trade, two-way paging is becoming the standard among people who prefer pagers. One unique application of PCS frequency based paging is a service which downloads a voice message to your pager; no more numeric, or text messaging to bother reading, listen to your message right from the pager. Figure 2.10 shows a typical two-way pager.

Wireless Data

Wireless data is the transfer of digital signals between two data devices via a wireless communication path. The growth in the wireless data marketplace in the late 1990's is comparable to the first stage of the cellular industry explosion in the early 1980's. Many established wireless service providers have committed (as well as major players in the computer industry) themselves over the next few years, to join in the wireless data revolution.

Figure 2.11 Wireless Email Pager
Source: Reasearch In Motion

With all the (new) wireless networks in place, a user simply needs to be informed of the new wireless "data-ready" applications that can help simplify life or generate new revenue.

Most wireless data services are dedicated to specific types of applications (vertical). Vertical wireless data applications are very specific solutions, and have continued to win over mass market "horizontal" offerings. Vertical solutions include applications such as utility meter reading or mobile dispatch. Horizontal solutions have mass-market appeal such as wireless e-mail shown in Figure 2.11.

Not until a mass-market wireless data application (often called the "Killer App") is embraced by the public will wireless data get the attention it deserves. Here are a few successful vertical wireless data applications.

Wireless data for the electric power, waste water and natural gas industries. New competition in the utility industry demands the benefits of a wireless data solution for timely customer focused improvements.

Wireless data for field service personnel. Field service organizations use wireless data to close the gap on a geographic distance to improve customer service, technician productivity and increased revenues.

Companies with mobile sales forces have increased their productivity and efficiency of personnel by filling out much of their paperwork "on-line". Sales force access to corporate databases has proven paramount in the new paradigm of doing business the 21st century style.

Wireless data transport technologies are discussed in Chapter 6. Again, based on the application, one certain technology may offer the best cost versus performance match for the user. As an example, if the application requires important bursts of data several times a day, a wireless data service offering designed for a permanent Internet connection would be too much. This example is looking for a wireless data service offering that promotes reliability and security, not speed or a permanent connection.

Fixed Wireless

Fixed wireless is a communications service between two devices that are fixed in location. The primary advantage of fixed wireless service is the ability to focus radio transmissions to a particular direction or region. This typically reduces interference to and from other radios and increases the capacity (data transfer rate) available to the fixed wireless device. The basic types of fixed wireless systems in use include wireless computer networks, competing wireless television systems, and wireless local telephone service.

Figure 2.12 Wireless LAN Product
Source: Proxim

Wireless Local Area Network (WLAN)

The most common fixed wireless application is the popular wireless local area network (WLAN) replacing common connections previously made by cable. Cable provides an excellent transmission media and supports data rates in the tens and hundreds of millions of bits per second. There are applications, however, that cannot use cable or are prohibitively expensive if cable is used.

Figure 2.12 shows product that are typically used in a WLAN system. This WLAN system includes radio access ports and extension ports. The extension ports shown in figure 2.12 are PCMCIA cards that plug into a laptop computer. These extension ports communicate via radio to radio access ports. The radio access ports convert the WLAN radio signal back into computer network signals (such as Ethernet or token ring).

Other WLAN applications, like point-of-sale terminals in the ever-changing retail environment make wireless access more cost-effective than cabled access. Mobile inventory scanning in warehouses tie WLANs to a wireless scanner. Some building architectures make cable installation prohibitively expensive, WLANs are well suited for these types of applications.

Often, Infrared (IR) light energy is used for point-to-point computer connections, because IR cannot pass through walls, ceilings, or floors. This is considered an advantage because it enhances the security of a WLAN link and decreases interference between other nearby WLANs.

Wireless Cable

Wireless Cable is the common term assigned to a radio frequency-based alternative to the cable TV distribution system. An example of wireless cable technologies is; multichannel multipoint distribution system (MMDS) or local multichannel distribution system (LMDS). Wireless cable can supply local television channels and high speed data services. MMDS has a high speed standardized air interface allowing mass deployment of cable television service by the new unregulated telephone companies. Cable television providers, who have access to most homes, can now provide telephone service.

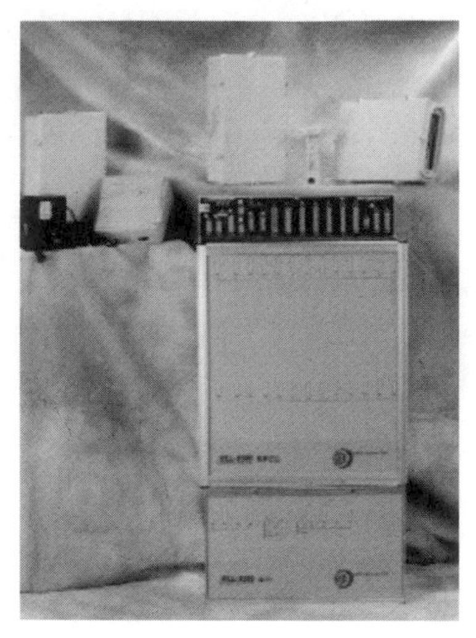

Figure 2.13 Wireless Local Loop
Source: World Access

Wireless Local Loop (WLL)

Wireless local loop (WLL) is the providing of local telephone service via radio transmission. WLL is primarily used in homes and businesses in developing countries because there is limited availability of wired telephone service. Because a majority of homes in the United States already have availability to wired telephone lines, wireless local loop service in the United States may be used to combine basic telephone service with high speed internet or video on demand (VOD) services. Providing wireless local telephone service also allows new competitive local exchange carriers (CLECs) to provide local telephone service without having to install large quantities of wires. Cable television providers, who have access to most homes, can also now provide telephone service. Figure 2.13 shows a wireless local loop receiver.

Remote Control

Remote control is the distant control of devices (commonly electronic or mechanical). Wireless remote control devices must conform to the FCC's part 15 unlicensed spectrum regulations. Typical uses of remote control devices can be as simple as a garage door opener shown in figure 2.14 or as complex as a Radio Remote Control System (RRCS) used for variable speed controllers on bomb-squad robots.

Figure 2.14 Garage Door Opener Remote Control
Source: Overhead Door

Remote controls are sophisticated because of the use of unlicensed frequency spectrum. The onus is on each remote control manufacturer to protect themselves against any intruding radio frequency energy.

Point to Point Communications Links

Point-to-point communication links is the transmission of information signals from one specific point to another, as distinguished from broadcast transmission, which blankets the information to the general public. A point-to-point connection is ordinarily a dedicated microwave radio channel that is transmitted between two antennas.

Dedicated point-to-point communication links are being used more than ever before for short distances. Microwave towers historically have been used to link distances of 50 miles at an expense quite considerably less than digging a ditch 50 miles long.

Figure 2.15 High Speed Wireless Data Microwave Communications Link
Source: Cylink

A recent allocation of frequencies in the 38 Giga-Hertz range has allowed for small, low cost wireless data microwave equipment to be utilized for short transmission links. Rather than lease a telephone circuit from the local telephone company, a Wireless Local Loop operator for example might link the cell site to the mobile switching center using a microwave wireless link. At microwave frequencies, many mega-bits of net data can be transported wirelessly point-to-point without the high costs associated with a wired connection. Figure 2.15 shows a high speed data microwave communication link.

Broadcast Radio

Radio broadcasting is the transmission of audio material (called a program) to a geographic area that is intended for general reception by the public, funded by air-time sold between programs.

AM and FM Audio Broadcasting

Amplitude modulation (AM) radio broadcast services have been available for the past 100 years. Most AM radio broadcast systems use relatively low radio frequencies and very narrow radio channel bandwidth to efficiently deliver audio information over large geographic areas. Unfortunately, low frequency used for AM transmission often result in signals that sometimes skip long distances (hundreds of kilometers). This has the potential for interference in distant cities. Amplitude modulation is also easily subject to electrical noise and signal distortion. Recent advancements in AM modulation can allow channel coding for stereo and more reliable (less distorted) radio signals.

To overcome some of the limitations of AM, frequency modulation (FM) was developed. FM transmission is less susceptible to noise and distortion. Unfortunately, most FM broadcast systems use a wider radio channel than AM systems. FM broadcast channels can be up to 20 times the bandwidth of a single AM broadcast channel. The latest advancements in FM broadcasting include conversion from analog to digital and the ability to simultaneously send some additional information (sub-channels) with their audio broadcasts (discussed below).

Sub Carrier Broadcasting

Two separate technologies are being tested to bring digital audio and data services to conventional radio broadcasts. The first, incorporates digital data into the conventional FM broadcast by adding the digital data signal to the existing audio signal before FM modulation. The second, is a fully digital transmission that is transmitted in addition to the conventional FM. This separate signal is added to the conventional FM signal after the FM modulation. Unlike High Definition Television (HDTV), these systems do not replace the analog service, they provide additional services and are completely compatible with conventional AM or FM broadcasts. The additional services are available only to those users with a receiver capable of accessing the digital data.

The entry of digital transmission into commercial broadcasting represents a revolution in the types of services that will be available to the public in the near future. Compare the possibilities to the many digital satellite features or the digital programming available with CD players. Imagine pressing one button on the car radio to request only news stations, or your preferred music category.

Digital Audio Broadcasting (DAB)

Digital audio broadcasting (DAB) transmits voice and other information using digital radio transmission. The DAB signal is normally shared with additional digital information on a single digital radio channel.

Broadcast Television

Television broadcasting combines the transmission of information typical to a telephone (sound) and information representing a visual image (VIDEO) Commercial television service started in the United States in 1946. By 1949, television receiver sales were exceeding 10,000 per month. In 1998, there were approximately 160 million television receivers in the United States.

Standard NTSC

The standard television system used in the United States is the national television standards committee (NTSC) system. The first version of this system was black and white television. The NTSC standard was later modified to allow color television signals to co-exist on the same type of video channel. The NTSC system is an analog system.

Several enhancements have been added to this basic system, including stereo encoding for the audio, additional audio programming channels, very low data rate digital transfer (closed captioning), and ghost canceling.

Television broadcasters transmit at high power levels from several hundred foot high towers. A high power television broadcast station can reach over 50 miles.

High Definition Television (HDTV)

The NTSC enhancements are minor, however; compared to the technological improvements represented by High Definition TV (HDTV) proposed to provide significantly higher resolution audio and video, as well as data services. A consortium called the Grand Alliance has produced a standard called Grand Alliance HDTV for digital television. The FCC plans to introduce HDTV initially by allowing broadcasters to offer a simulcast of their regular programming, transmitted on UHF television assignments. The period of simulcast will continue for

Figure 2.16 High Definition Television
Source: Toshiba

up to 15 years as old broadcast facilities and receivers are phased out. Receivers for the HDTV system will also include the capability to receive and display regular analog broadcasts.

Chapter 3

Wireless Basics

Wireless communication involves the transfer of information signals through the air by the means of electromagnetic waves. To create electromagnetic waves, an electrical signal, that continuously varies in power level and polarity, is applied to an antenna. As the level varies, the energy contained in the electrical signal is converted to electromagnetic waves that propagate away from the antenna. The electromagnetic waves are characterized by their energy (RF power) and frequency (cycles per second or Hz named after the German physicists Heinrich Rudoph Hertz). Commercial uses for these electromagnetic waves repeat their cycle in a frequency range of approximately 150,000 Hertz (150 kHz) to 300 billion (giga) Hertz (300 GHz).

To allow information to be transferred using electromagnetic waves, an information signal (typically audio) slightly changes the wave shape of the electromagnetic signal. This is called a radio signal.

Radio signals can co-exist with each other without interference if they are operating at different frequencies. Because an information signal slightly changes the electromagnetic signal, this produces small changes in frequency. This results in a single radio signal that occupies a frequency range, depending on the type and amount of information that is changing the electromagnetic wave. The maximum amount of frequency change is commonly called the channel bandwidth. Hence, a radio signal should not ordinarily operate in areas that other radio signals may occupy.

Dividing the Radio Frequency (RF) Spectrum

The radio frequency spectrum is divided into frequency bands that are authorized for use in specific geographic regions. In the United States, responsibility for allocating the radio wave portion of our electromagnetic spectrum belongs to the Federal Communications Commission (FCC). The government's interest in this responsibility is multi faceted; in addition to United States defense interests, the FCC's primary activities and decisions are for the benefit of the general public.

The FCC began by defining the highest level of frequency division. Table 3.1 shows the typical categorization of frequency bands and types of use. The most commonly recognized names are the two associated with television channels; VHF (very high frequency) and UHF (ultra high frequency) where older televisions had two dials; one dial was VHF and the other UHF.

Frequency Range	*FCC Naming Convention*	*Common Name*	*Example Use*
below 3 kHz	Extremely Low Frequency	ELF	under water
3 kHz to 30 kHz	Very Low Frequency	VLF	Navy
30 kHz to 300 kHz	Low Frequency	LF	Maritime
300 kHz to 3 MHz	Medium Frequency	MF	AM Radio
3 MHz to 30 MHz	High Frequency	HF	FM Radio
30 MHz to 300 MHz	Very High Frequency	VHF	Paging
300 MHz to 3 GHz	Ultra High Frequency	UHF	Cellular
3 GHz to 30 GHz	Super High Frequency	SHF	Microwave
30 GHz and above	Extremely High Frequency	EHF	Satellite
3x1014 Hz	Infrared Frequency	IR	WLAN
1x1016 Hz	Visible Light Frequency	Lightwave	Fiber optics

Table 3.1 Frequency Bands and Their Use

Licensing

The FCC is responsible for dividing the available frequency bands for licensing to users and regulates what the frequencies may be used for. The legal right-to-use of this public resource is controlled by rules and licensing of very specific frequencies, a range of frequencies or a block of sub-divided channels at a given frequency or frequency range.

For example, the frequencies allocated for FM radio must be used for the purpose licensed; that is a combination of music or news and public information. FM radio stations are not licensed to broadcast a secret "Morse-code" to a following of undercover militia! Neither can a "Paging Service" use one or all of their frequency channels to broadcast radio. However, with the recent deregulation of telecommunications services, wireless service providers are now permitted to offer many new types of services provided they can fulfill their basic licensing requirements.

To prevent unwanted interference from radio devices, the reckless use of transmitting energy or information on our public airwaves according to publicly published rules or licenses will violate federal law. Such transmissions are subject to prosecution or suspension of the radio operator's license.

Frequency Allocation Charting

There are hundreds of wireless applications that are assigned many different frequency bands. The selection of the assigned frequency bands was determined by a variety of factors including the radio propagation characteristics, availability of radio channel frequencies at the time and the determination of the FCC that the use of the radio frequency would meet the needs of the public.

Because most of the frequencies have already been assigned to licensees, a new assignment of frequencies normally requires existing licensees or users to stop using a band. These users are regularly shifted to another band. This process is called re-allocation.

Historically, major re-allocations are done in the higher frequencies to avoid congestion. This has advantages and disadvantages. The radio frequency (RF) devices employed within the newer systems are subject to more loss based on distance. This requires closer distances, increasing the total number of radio sites to cover the same area previously covered by radio devices at a lower frequency. However, the higher frequencies tend to penetrate buildings more readily and the antennas involved are physically smaller - both important attributes for systems that seek to reach 100% of the available population.

RF Channels and Bandwidth

An RF channel is a communication link that uses radio signals to transfer information between two points. To transfer this information, a radio wave (usually called a radio carrier) is modulated (modified) within an authorized frequency band to carry the information. The modulation of the radio wave forces the radio frequency to shift above and below the reference (center) frequency. As a rule, the more the modification of frequency, the more information can be carried on the radio wave. This results in RF channels being defined by their frequency and bandwidth allocation.

Bandwidth allocation is the frequency width of a radio channel in Hertz (high and low limits) that can be modulated to transfer information.. The amount of bandwidth used is determined by the amount and type of information being sent, as well as the method of modulation used to impose the information on the radio signal.

The FCC defines a total frequency range (upper and lower frequency limits) that a radio service provider can use to transmit information in a defined geographic area. In some systems (such as AM or FM radio station broadcasting), this is a single radio channel. For other systems (such as cellular or PCS), this is a range of frequencies that can be sub-divided into smaller radio channels as determined by the radio carrier. When the allocated frequency range is further subdivided into smaller allowable bands, these subdivided areas are referred to as channels.

Signal Types

There are two major types of signals; analog and digital. Until recently, most radio systems exclusively used analog signals. Over the past ten years, many systems have started using digital radio signals.

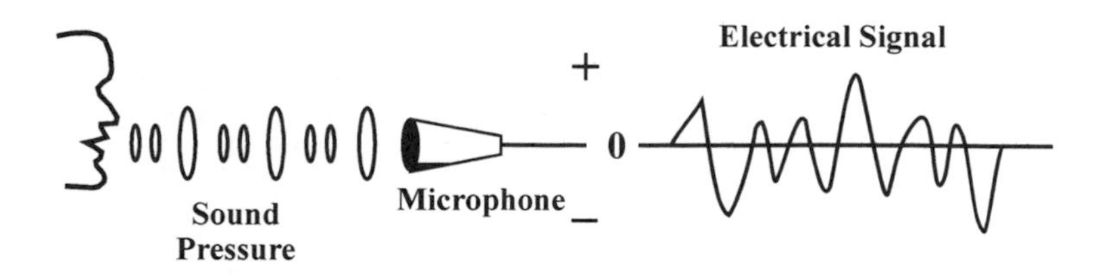

Figure 3.1 Analog Audio Signal

Analog

An analog signal is a signal that can vary continuously between a maximum and minimum value, like a dimmer switch found in many household dining rooms. An analog signal can assume an infinite number of values between the two extremes.

Figure 3.1 shows a sample analog signal created by sound. In this example, as the sound pressure from a person is detected by a microphone, it is converted to its equivalent electrical signal. Notice that, as shown in Figure 3.1, the analog audio signal continuously varies in amplitude (height, loudness, or energy) as time progresses.

Digital

Digital refers to a signal or category of electronic devices that represent information by discrete signal levels that change at predetermined intervals. Digital signals typically have two levels; on (logic 1) and off (logic 0). The information contained in a single time period is called a bit. The number of bits that are transferred in one second is called the data transfer rate or bits per second (bps). Because many bits are commonly transferred in one second, the bps designation of the data rate is ordinarily preceded by a multiplier k (thousand) or M (million). For example, if the data transfer rate is 3 million bits per second, this would be indicated by 3 Mbps.

Bits are normally combined into groups of 8 bits to form a byte. When the reference is made to bytes instead of bits, the b is capitalized. For example, 10 thousand bytes is represented by kB. Figure 3.2 shows a sample digital signal. In this example, the bits 01011010 are transferred in 1 second. This results in a bit rate of 8 bps.

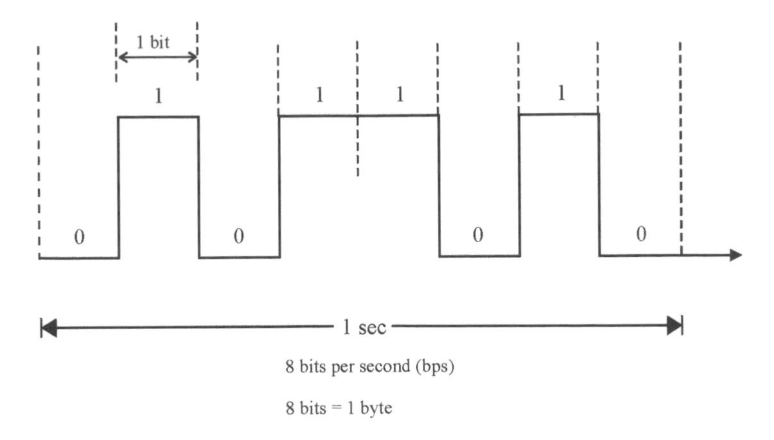

Figure 3.2 Digital Signal

The earliest form of digital radio communication was Morse Code. To send Morse Code, the radio transmitter was simply turned on and off to form dots and dashes. The rcceiver would sense (detect) the radio carrier to reproduce the dots and dashes. A code book of dots and dashes was used to decode the message into symbols or letters. The on and off pulses or bits that comprise a modern digital signal are sent in a similar way.

The trend in wireless systems, just as in other types of electronics products such as compact discs, is to change from analog systems to digital systems. Digital systems have a number of important advantages including the fact that digital signals are more immune to noise and, unlike analog systems, even when noise has been introduced, any resulting errors in the digital bit stream can be detected and corrected. Also, digital signals can be easily manipulated or processed in useful ways using modern computer techniques.

Digitization of an Analog Signal

Analog signals must be converted to digital form for use in a digital wireless system. The analog signal is digitized (converted to a digital signal) by using an analog-to-digital (A/D - pronounced A to D) converter. The A/D converter periodically senses (samples) the level of the analog signal and creates a binary number or series of digital pulses that represent the level of the signal. This technique is called Pulse Code Modulation. The typical sampling rate for voice signals occurs at 8000 times a second. Each sample produces 8 bits of digital informa-

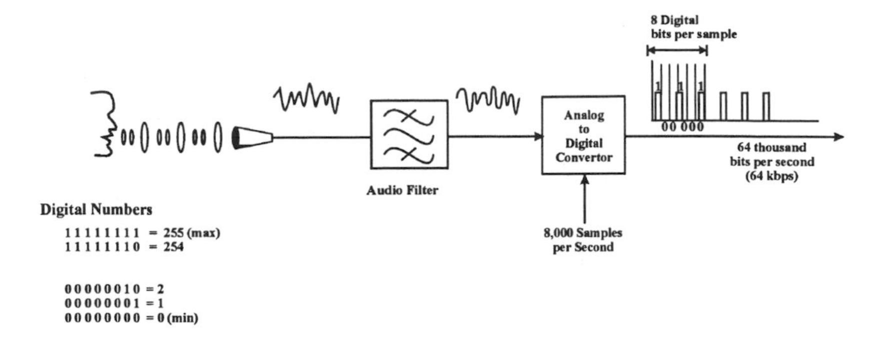

Figure 3.3 Signal Digitization

tion which results in a digital data rate (bit stream) of 64 thousand bits per second (kbps). Figure 3.3 shows how an analog signal is converted to a digital signal.

Many systems that have traditionally used analog signals are now being converted to systems that use digital signals. Digital signals allow for more reliable radio transmission, improved quality, capacity and flexibility.

Digital signals are more reliable because of their resistance to noise and the ability to perform error detection and correction. Digital signals have two discrete levels (1 and 0). Small amounts of noise or interference do not affect these levels. Digital signals often contain additional data bits that allow for the detection and correction of bits that have been received in error.

Modulation of the Radio Waves

Radio modulation is the process of modifying the characteristics of a radio carrier wave (electromagnetic wave) using an information signal (such as voice or data). The characteristics that can be changed include the amplitude modulation (AM), frequency modulation (FM), or phase modulation (PM). A pure radio carrier signal carries no information. When the radio signal is modified from a normalized state, it is called a modulated signal (thus containing information). This modulated signal is the carrier of the information that is used to modify the carrier signal. When the radio carrier is received, its signal is compared to an unmodulated signal to reverse the process (called demodulation). This allows the extraction of the original information signal.

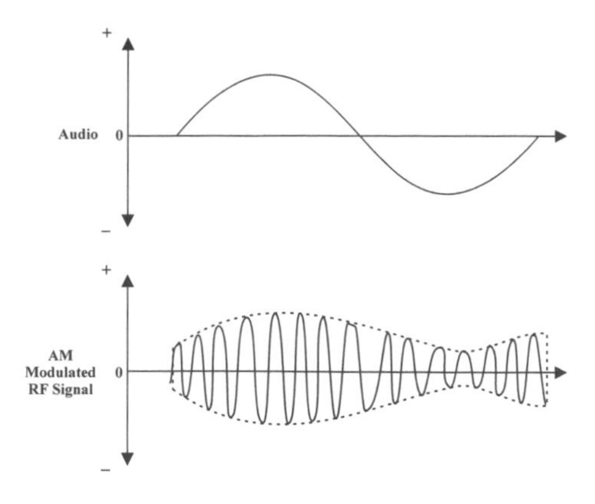

Figure 3.4 Amplitude Modulation

Amplitude Modulation (AM)

Amplitude modulation is the transferring of information onto a radio wave by varying the amplitude (intensity) of the radio carrier signal. AM is the simplest form of modulation. Figure 3.4 shows an example of an AM modulated radio signal (on bottom) where the height of the radio carrier signal is changed by using the signal amplitude or voltage of the audio signal (on top).

Frequency Modulation (FM)

In 1936, Armstrong demonstrated an FM transmission system that was much less susceptible to noise signals than AM modulation systems. Frequency modulation involves the transferring of information onto a radio wave by varying the instantaneous frequency of the radio carrier signal. Figure 3.5 illustrates a process known as frequency modulation (FM). In this diagram, as the modulation signal (audio wave) increases in voltage, the frequency of the radio carrier signal increases. As the voltage decreases, the frequency of the radio carrier signal decreases.

One form of frequency modulation used to transmit digital information is called frequency shift keying (FSK). To represent a digital signal, the FSK modulator transmits on one frequency to signify a one (on) and a different frequency to signify a zero (off).

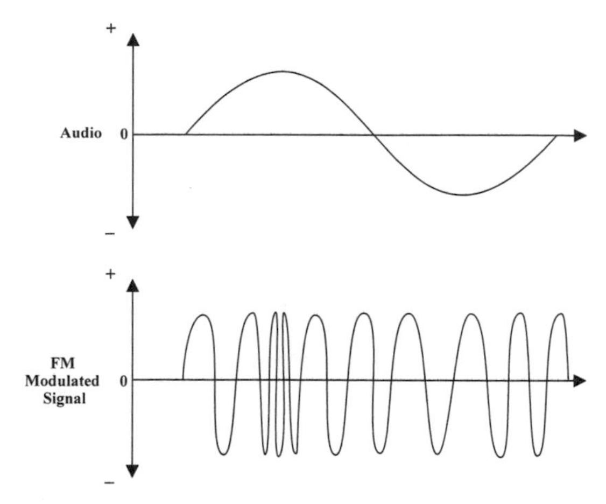

Figure 3.5 Frequency Modulation

Phase Modulation (PM)

Phase modulation is a modulation process where the phase of the radio carrier signal is modified by the amplitude of the information (typically audio) signal. Changes from an input source, reflected by correspondingly varying the phase (or relative timing) of the carrier wave signal as shown in Figure 3.6, which shows a sample of phase modulation (PM) with a digital input signal. In this diagram, the digital signal (on top) creates a phase modulated RF signal (on bottom). As the digital signal voltage is increased, the frequency of the radio signal changes briefly so the phase (relative timing) of the transmitted signal advances compared to the unmodulated radio carrier signal. This results in a phase shifted signal (solid line) compared to an unmodulated reference radio signal (dashed lines). When the voltage of the digital signal is decreased, the frequency changes again so the phase of the transmitted signal retards compared to the unmodulated radio carrier signal.

Today's sophisticated modulation systems can use all three variable parameters: frequency, amplitude or timing (phase) at the same time to transfer analog or digital information.

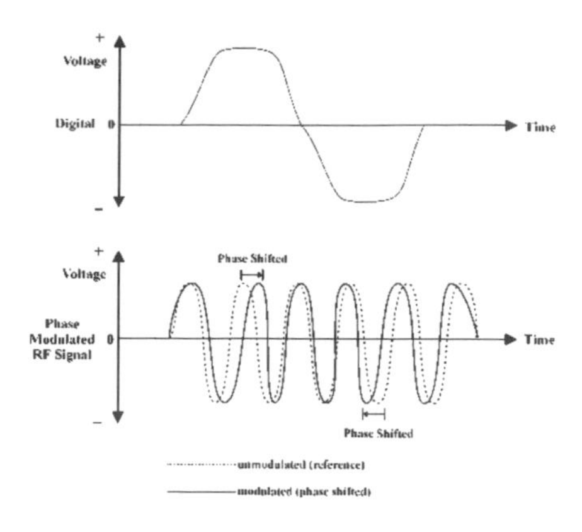

Figure 3.6 Phase Modulation

System Access Technologies

The method used by mobile radio systems to coordinate service requests and share radio channel usage is through the use of system access technologies. System access technologies are the language of radio. Mobile radios can communicate with a system, as long as the system can understand their language. System access technologies are commonly defined by their modulation type, timing and information characteristics. There are three basic ways that allow radios to share access to radio channels; frequency division, time division and code division multiplexing.

There are two basic functions for radio channels used in access systems: control and communication. Control involves the coordination between transmitters and receivers so they can request service and avoid interference with each other. Communication involves the transfer of information content (voice or data) between two points. These channels may allow only one user at a time to send information or they may be divided to allow the sharing (multiplexing) of the radio channel by multiple radios at the same time.

When a radio channel is exclusively used to coordinate access to other radio channels, it is called a control channel or a signaling channel. When a radio channel is used to transfer information between radios, it is called a voice or traffic channel. Some radio access technologies combine control and communication on the same RF channel.

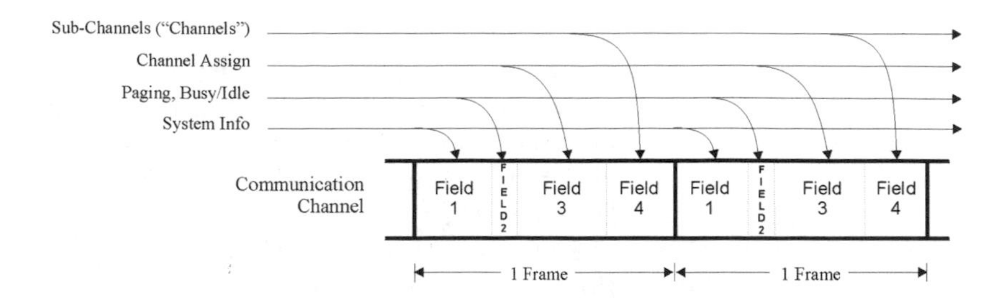

Figure 3.7 Radio Channel Fields

Each radio communication channel is typically divided into several sub channels. The radio channel is typically called a "communication channel" and each sub-channel is called a "channel." To create sub channels, information sent on the radio channel is grouped into small sections called fields. Fields are grouped into frames.

Figure 3.7 shows how a single radio channel is divided into fields and frames to create sub channels. The sub channels in this example contain system information, system status (idle/busy), paging and channel assignment. A radio listens to the radio channel to capture the information contained in one or more of the fields.

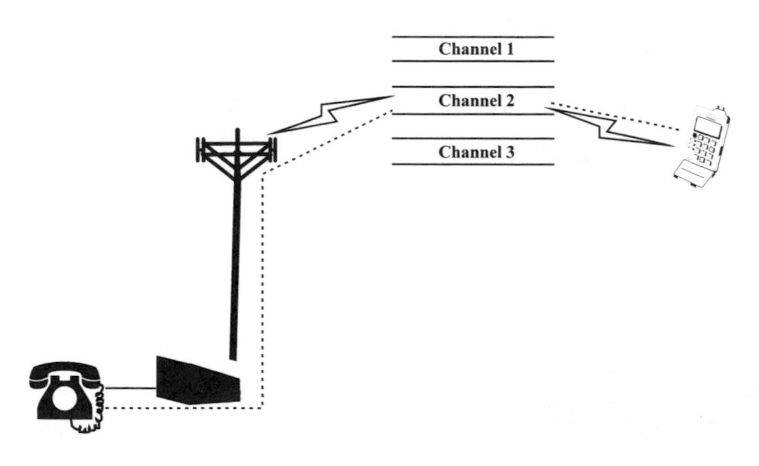

Figure 3.8 Frequency Division Multiple Access

Frequency Division Multiple Access (FDMA)

Frequency division multiple access is a process of allowing mobile radios to share radio frequency allocation by dividing up that allocation into separate radio channels where each radio device can communicate on a single radio channel during communication. Frequency division multiple access (FDMA) was the first access technology used for two way radios. Figure 3.8 shows how several radio channels can be accessed by a mobile radio. When a mobile radio is communicating with the system, its radio channel is completely occupied by the radio transmission of the mobile radio. That communication channel is busy until after the mobile radio has completed its communication. After it has stopped transmitting, other mobile radios can be assigned to that radio channel frequency. Mobile radios in an FDMA system typically have the ability to tune to several different radio channel frequencies.

Time Division Multiple Access (TDMA)

Time division multiple access (TDMA) is a process of sharing a single radio channel by dividing the channel into time slots that are shared between simultaneous users of the radio channel. When a mobile radio communicates with a TDMA system, it is assigned a specific time position on the radio channel. By allowing several users to use different time positions (time slots) on a single radio channel, TDMA

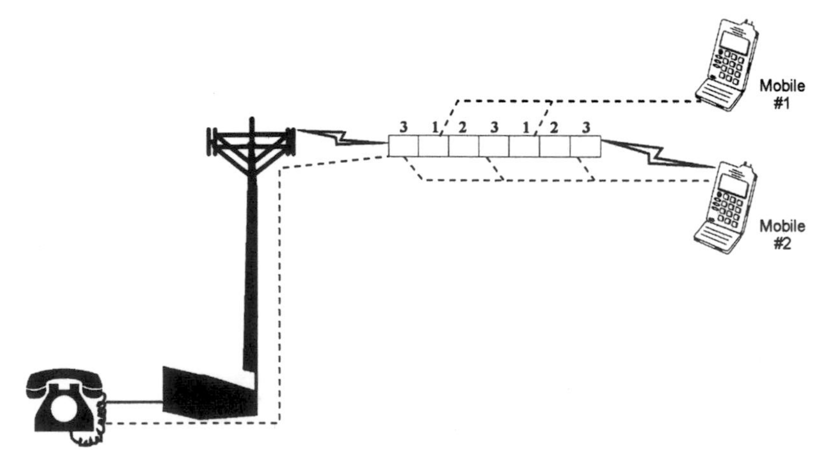

Figure 3.9 Time Division Multiple Access

systems increase their ability to serve multiple users with a limited number of radio channels.

To allow TDMA systems to provide continuous voice communication to a radio that can only transmit for brief periods, TDMA systems use digital signal processing to characterize and compress digital signals into short time-slices. Figure 3.9 shows how a single radio channel is time-sliced into three communication channels. Mobile radio number 1 is communicating on time slot number 1 and mobile radio number 2 is communicating on time slot number 3.

Spread Spectrum (FHMA and CDMA)

Spread spectrum communications is a method of spreading information signals (typically digital signals) so the frequency bandwidth of the radio channel is much larger than the original information bandwidth. There are various forms of spread spectrum communications. The most popular forms of spread spectrum include frequency hopping multiple access (FHMA) and code division multiple access (CDMA).

FHMA is an access technology where mobile radios may share radio channels by transmitting for brief periods of time on a single radio channel and then hopping to other radio channels to continue transmission. Each mobile radio is assigned a particular hopping pattern and collisions that occur are random and only cause a loss of small

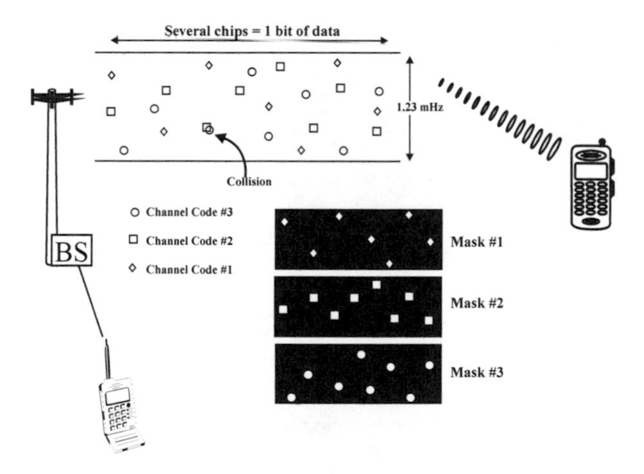

Figure 3.10 Code Division Multiple Access

amounts of data that may be fixed through error detection and correction methods.

CDMA is a relatively new commercialized (verses militarized) modulation technique that is used in cellular and satellite systems. CDMA systems mix a relatively long digital code with a small amount of communication data to produce a combined signal that is spread over a relatively wide frequency band. To receive the signal, the long code is used to extract the original signal.

Because the energy is spread over a wide bandwidth, multiple CDMA channels with different codes can co-exist with minimal interference. Figure 3.10 shows how a single CDMA radio channel can have several channels. In this example, there are three different code patterns that are used for communication channels. Knowing the code, a CDMA system can build a mask as shown in Figure 3.10 for each conversation allowing only that information which falls within the mask to be transmitted or received.

Duplex Operation

Mobile radio systems typically require two way communications where both users can communicate at the same time. To allow two way communications, mobile systems are either frequency division duplex (FDD), time division duplex (TDD) or combined FDD and TDD. TDD systems can use FDMA, TDMA or CDMA access technologies.

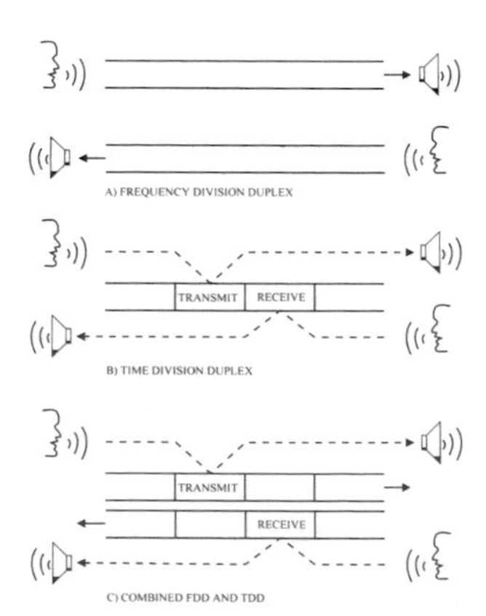

Figure 3.11 Duplex Systems

Frequency division duplex (FDD) systems use two frequencies to allow simultaneous communication. One frequency is used to communicate in one direction and the other frequency is required to communicate in the opposite direction. Figure 3.11 (a) shows FDD operation.

Time division duplex (TDD) communication uses a single frequency to provide two way communications between devices by time-sharing. When using TDD, one device transmits (device 1), the other device listens (device 2) for a short period of time (usually within 100 ms). After the transmission is complete, the devices reverse their role so device 1 becomes a receiver and device 2 becomes a transmitter. The process continually repeats itself so data appears to flow in both directions simultaneously. Figure 3.11 (b) shows a TDD system operation.

Figure 3.11 (c) shows a system that combines FDD and TDD operation. Combining FDD and TDD offers the benefit of simplified radio design.

Analog (Audio) Signal Processing

Analog signals may be processed by filters, shaping circuits, combiners and amplifiers to change their shape and modify their content. Audio signals contain a wide range of frequencies that are not necessary to transmit. These high and low frequencies can be removed by filtering. Audio signals may be processed by shaping circuits that add or remove emphasis to frequency (tone) or intensity (volume). When

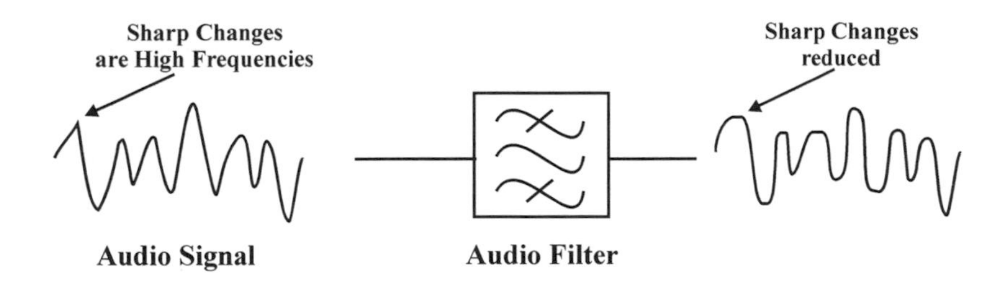

Figure 3.12 Audio Signal Processing

analog audio signals are used to modulate a radio wave, the high frequency components of the audio signal are emphasized (called pre-emphasis) to allow the modulator to be more effective. Because the intensity of an audio signal can vary dramatically (some people talk loudly and others talk softly), an amplification of the audio signal may be reduced as the user talks louder (called companding). When pre-emphasis and companding are used to process an audio signal in a transmitter, their opposite functions (de-emphasis and expanding) are used in the receiver to restore the audio signal back to its original form.

In some cases, additional signals may be combined with audio signals prior to transmission. These signals are used for control purposes. If control signals are added to an audio signal in a transmitter, they must be removed from the audio signal in the receiver by filtering. Audio amplifiers are used to adjust the volume of an audio signal.

Figure 3.12 shows typical audio signal processing for an analog radio transmitter. In this example, the audio signal is processed through a filter to remove very high and very low frequency parts. The high frequencies can be seen as rapid changes in the audio signal. After the audio signal is processed by the filter, the sharp edges (high frequency components) are removed.

Digital Signal Processing

Digital signals are manipulated by microprocessors (limited computers) to change their content and to add error detection and correction capability. When analog signals are converted to digital format, the digital signals represent the original analog waveform. Just like analog signals that may be processed by filters, shaping circuits, combiners and amplifiers, digital signals can be processed to produce similar functions.

To change a digital signal, a microprocessor is used to manipulate the incoming digital information via a program (stored instructions) that produce a new digital output. The program determines the functions that are performed in the digital signal.

Because radio signals can experience signal distortion that can result in the incorrect determination of a digital signal (whether a zero or one had been sent), digital systems typically include error detection and correction processing. Error detection processing involves the creation of additional bits that are sent with the original data. The check bits are created by using a formula calculation on the digital signal prior to sending. After the digital signal is received, the formula can be used again to create check bits from the received digital signal. If the check bits match, the original digital signal was received correctly. If the check bits do not match, some or all of the digital signal was received in error. This process is called error detection.

Some digital systems use sophisticated mathematical formulas to create the check bits so that the check bits can be used to make corrections (or predictions of the correct bits) to the received digital signal. This process is called error correction.

Early communication systems contained one or more microprocessors along with all other integrated circuits (such as memory and interface circuits) to perform digital signal processing. To reduce the number of components in a digital radio, digital signal processors (DSPs) and custom application specific integrated circuits (ASIC) are typically used. DSPs and ASICs are specialized integrated circuits that combine many electronic circuits and sometimes even contain the programs that process digital signals.

Figure 3.13 shows typical digital signal processing that is used in a digital radio system. In this diagram, a 64 kbps digital signal that represents an audio signal is processed (compressed) to form a 8 kbps signal. This process is called speech coding. The microprocessor uses a stored program for its instructions that analyze the signal so it can convert the original high speed digital signal into a lower speed signal. As the signal is analyzed, it is compared to numbers that closely

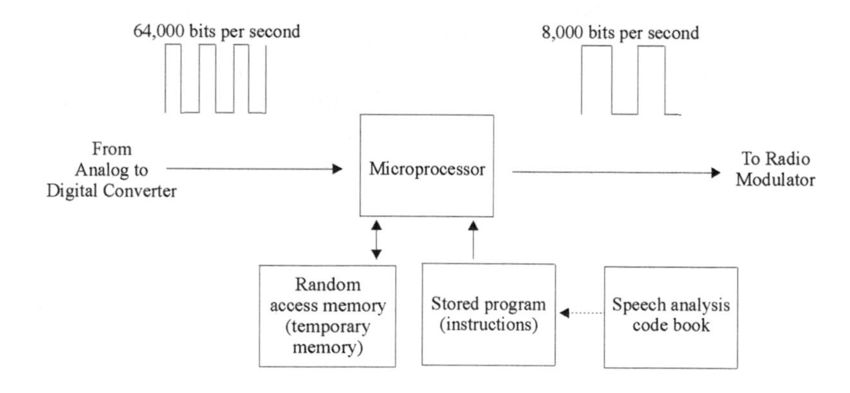

Figure 3.13 Digital Signal Processing

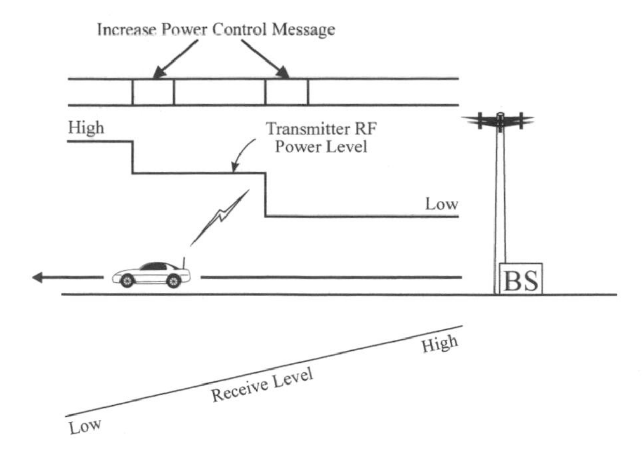

Figure 3.14 RF Power Control

match it in the code book. The codes replace the original digital signal. After it is processed, the digital signal is sent to the radio modulator.

Radio Signal Processing

RF signal processing includes RF amplification and filtering. RF amplification is used to increase (amplify) a radio signal to a high level RF signal for transmission or to boost the low level of a received RF signal. RF filtering allows the passing of desired radio signals (such as the radio channel the receiver is listening to) while blocking unwanted signals (such as other radio channels).

The cellular system typically controls the transmitter power level of mobile phones so they do not interfere with other nearby mobile telephones or radio towers. The cellular system typically senses the RF signal that is received from the mobile phone. This level is used to determine the correct transmitter power level the mobile phone should be adjusted to. The cellular system then sends commands to the mobile phone to adjust the power level accordingly. Figure 3.14 shows how the system sends commands to a mobile phone to increase its power level as it moves away from the base station.

Antennas

Antennas convert electrical signals into electromagnetic waves that travel through the air. Antennas must be designed specifically according to the frequency of the application. The most obvious design parameter for an antenna is the size or length of the antenna. All antennas have an optimal relation to the size of one radio wave (wavelength)of the frequency in question. The higher the frequency the smaller the wavelength, and the smaller the antenna can be. At 800 MHz (the cellular frequency) the natural wavelength is about 15 inches. A popular cellular antenna size is the ¼ wave antenna. This results in a cellular antenna being approximately 9 cm (3 ½ inches) long.

Gain

While antennas by themselves cannot add energy (amplify) a radio signal, antennas can provide signal gain by focusing the radio signal in a particular direction. This is possible by using the energy that would have been transmitted in other directions. Antenna gain is the amount of energy that is transmitted from the antenna in a particular direction compared to the input energy level.

It is not always desirable to have antenna gain. For example, handheld telephones typically have only a small amount of antenna gain so they can transmit equally in most directions. Figure 3.15 shows typical types of antenna gain.

The gain of an antenna is typically expressed in decibels (dB). Decibels are 1/10th of a Bell (hence the prefix deci-). The measurement of Bells is the logarithmic ratio of the input signal level with the output signal level. A single Bell is 10 times the power level from input to output. Because 1 Bell equals 10 decibels, 10 dB is also 10 times the power level. Decibels are logarithmic which means the measurement of bells is multiplied. For example, a gain of 20 dB is a power gain of 100 times. A popular amount of gain for mobile telephone antennas is 3 dB, about 2 times the amount of power in the focused direction.

Radios

Radios may be fixed in location (such as a television) or may be mobile (such as a cellular telephone). Some radios may only communicate in one direction (typically a receiver) or may have two-way capability. When a single radio has both a transmitter and receiver contained in the same unit, it is called a transceiver.

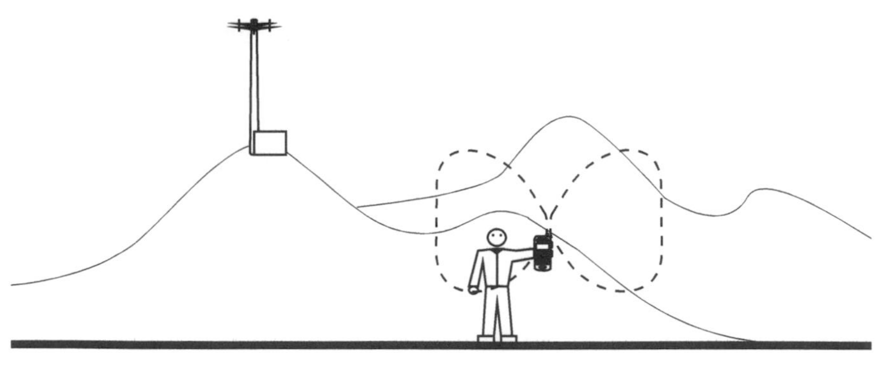

A) 1 dB Gain

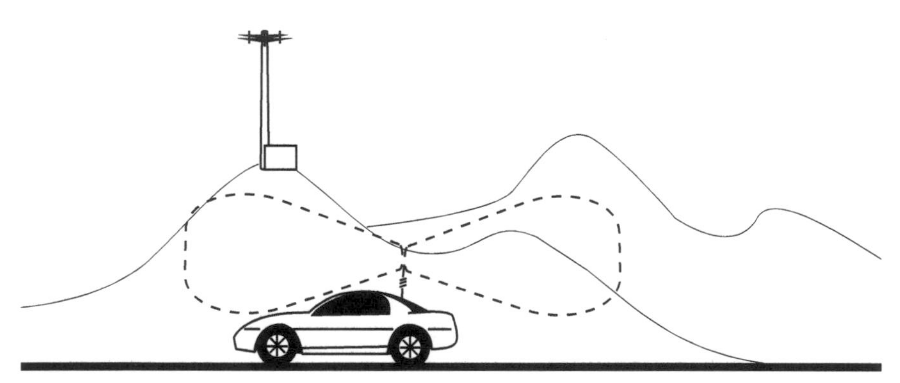

B) 3 dB Gain

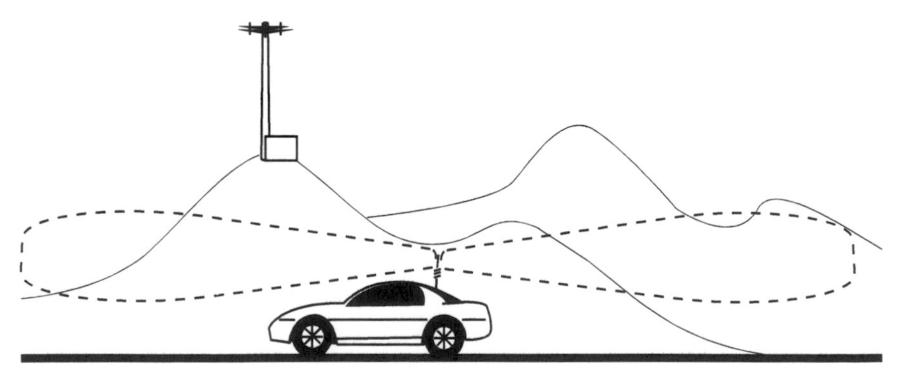

C) 5 dB Gain

Figure 3.15 Antenna Gain

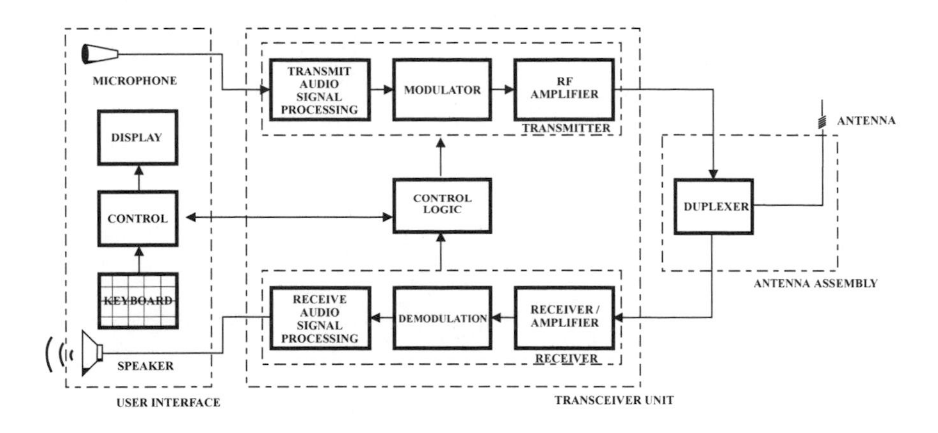

Figure 3.16 Mobile Radio Block Diagram

Figure 3.16 shows a block diagram of a mobile radio transceiver. In this diagram, sound is converted to an electrical signal by a microphone. The audio signal is processed (filtered and adjusted) and is sent to a modulator. The modulator creates a modulated RF signal using the audio signal. The modulated signal is supplied to an RF amplifier which increases the level of the RF signal and supplies it to the antenna for radio transmission. This mobile radio simultaneously receives another RF signal on a different frequency to allow the listening of the other person while talking. The received RF signal is boosted by the receiver so it can be applied to the demodulator. The demodulator extracts the audio signal and the audio signal is amplified so it can create sound from the speaker.

Wireless Network Infrastructure

Radio Towers

Radio towers are poles, guided towers or free standing constructed grids that raise one or more antennas to a height that increases the range of a transmitted signal. Radio towers can vary in height from about 20 feet to more than 300 feet. A single radio tower may host several antenna systems that include paging, microwave or cellular systems. Radio towers are located strategically around the city to provide radio signal coverage to specific areas. At the base of the towers are electronic control rooms which contain the components to operate the radio portion of the communications system.

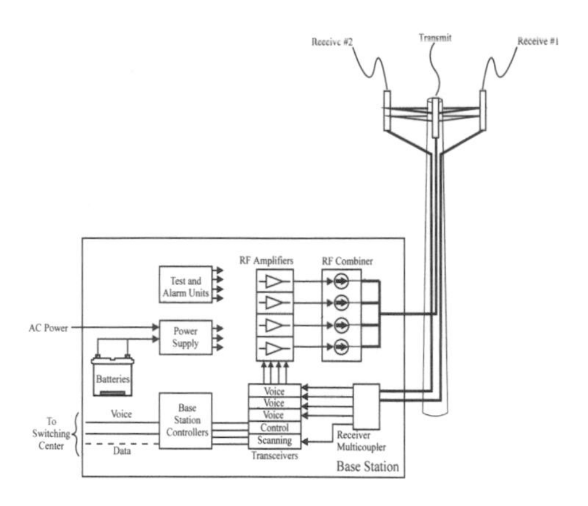

Figure 3.17 Radio Base Station Block Diagram

Base stations consist of major components like those shown in Figure 3.17. These may include one or more antennas, transmitters, receivers (for two-way systems), system controllers, communication links and power supplies. Transmitters provide the high level RF power that is supplied to the antenna. For broadcast systems, the amount of transmitter power can exceed 50,000 Watts. Receivers boost and demodulate incoming RF signals from mobile radios. If a base station contains receivers, it is typical to use one or more different antennas for the receivers. Controllers coordinate the overall operation of the base station and coordinate the alarm monitoring of electronic assemblies. Communication links allow a command location (such as a television studio or a telephone switching center) to control and exchange information with the base station. Base station radio equipment requires power supplies. Most base stations contain primary and backup power supplies. A battery typically maintains operation when primary power is interrupted. A generator may also be included to allow operation during extended power outages. Figure 3.17 shows a typical radio base station block diagram.

Switching Facilities

Switching facilities are typically used in two-way mobile communication systems to allow the connection of mobile radios to other radios in the system or to the public telephone network. When used in a cellular system, the switching system is typically called a Mobile Switching Center (MSC). The MSC, just like a local telephone company, process-

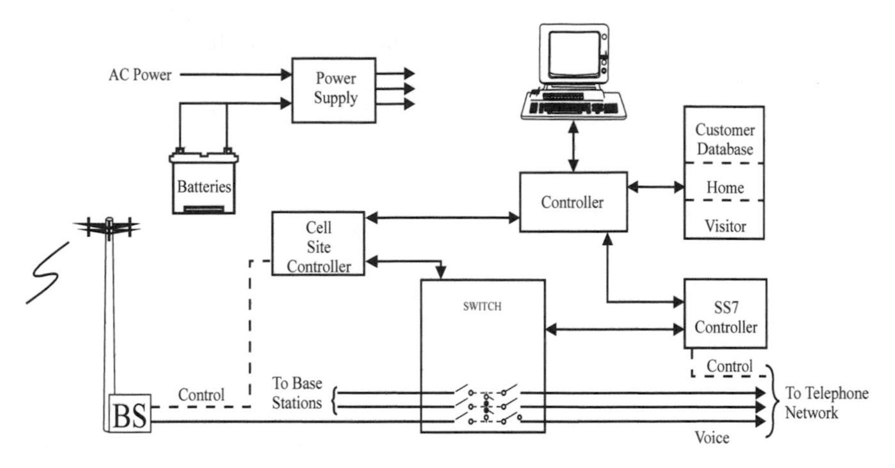

Figure 3.18 Wireless Switching System

es requests for service from mobile radios (subscribers) and routes the calls to other destinations.

Figure 3.18 illustrates a wireless switching system basic functional components: a customer database, system and communication controllers, a switch, primary and backup (batteries) power and the interface to the radio tower's base station (BS).

Interconnection to Other Networks

Almost all wireless systems are connected to other networks. Broadcast systems may be connected to media sources (such as audio or video programs) via satellite links while cellular networks may be interconnected to the public telephone network. Interconnection involves the physical and software connection of network equipment or communications systems to the facilities of another network such as the public telephone network. Government agencies such as the Federal Communications Commission (FCC) or department of communications (DOC) regulate interconnection of wireless systems to the public telephone networks to ensure reliable operation.

Telephone Numbering

Mobile systems have a key challenge that wired telecommunications systems do not. Wireless networks must monitor the location of wireless users in its system and other systems the user is visiting, to determine where to alert the user that calls are coming to them. To allow the tracking of mobile users, several telephone signaling systems are used to locate the user. When the user is found, the telephone call must be re-routed to the telephone number of the system or location where the mobile phone is operating.

Customer Databases

Customer databases are computer storage devices (typically a computer hard disk) that hold service authorization and feature preferences of customers. Each wireless subscriber has a real-time user profile in the database that is typically called the Home Location Register (HLR). The HLR identifies the current location of the mobile radio, the most likely place for the mobile to be, or the last location the subscriber was active. The MSC system controller uses this information to route calls to the appropriate radio tower for call completion. If the wireless user is not in a predetermined "home" range of the MSC, the mobile will register back through to the home signaling system to its Home Location Register (HLR) for profile information. This information will be temporarily stored in a Visitor Location Register (VLR). Within the past few years, Signaling System 7 (SS7) has been deployed in wireless communications which makes roaming much easier than ever before. A cellular subscriber can be located by simply dialing their home number regardless of their physical locations.

System Security

In some wireless networks, access to system services requires validation of the customers identity. These systems may use an Authentication Center (AUC) to store and process secret data to stop fraudulent calls or prohibit access to other paid for subscription services.

Wireless phones transmit some of their identification information over the public airwaves when they attempt to access the system. Thieves may try and intercept this information and copy (clone) the identification information that would allow them to make phone calls that would be billed to the other telephone. To prevent this unauthorized duplication of identification information, an authentication process can be used that uses secret keys to validate access information.

During the authentication process, code keys are created from secret codes that are stored in both the mobile radio and in the system. Along with basic identification information, these keys are exchanged during each system access attempt. The secret codes are not transmitted. Because the system and the mobile radio have the secret keys, both the mobile phone and the system can validate that the code information is correct. If the codes do not match, the system should not allow the call to be processed. New codes are created during each access attempt to prevent copying of the codes and immediately attempting access.

Public Switched Telephone Network

The Public Switched Telephone Network (PSTN) is the common wired telephone system used by the world to connect any telephone to any other telephone connected to the PSTN. Figure 3.19 is a basic overview of the PSTN system. The parts of the PSTN system are interconnected by different levels of switches. Lower level switches are used to connect end users (telephones) directly to other end users in a specific geographic area. Higher level switches are used to interconnect lower level switches.

Switches send control messages to each other through a separate control signaling network called signaling system number 7 (SS7). The SS7 network is composed of signaling transfer points (STPs) and service control point (SCP) databases. A STP is used to route packets of control messages through the network. SCPs are databases that are

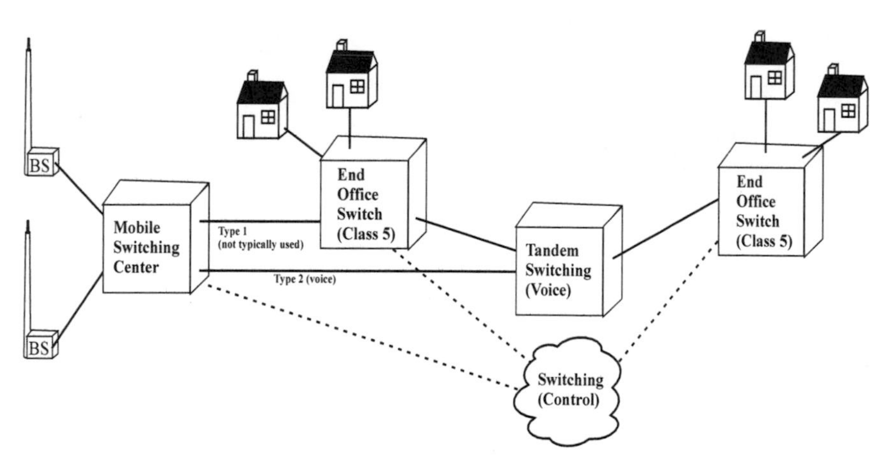

Figure 3.19 Public Switched Telephone Network

used by the network to process or reroute calls through the network (such as 800 number toll free call routing).

Figure 3.19 also shows how a mobile wireless network may connect to the PSTN. Two types of connections are shown: voice and signaling. End office (EO) and tandem office (TO) switches route voice connections. The end office (EO) switch is nearest to the traditional customer telephone. The Tandem office (TO) switching systems connect end office (EO) switches when direct connection to an end office is not justified. Tandem office switches can be connected to other tandem office switches.

Many countries have started to separate telephone networks into local exchange carriers (LECs) and an inter-exchange carriers (IXCs). LEC providers furnish local telephone service to end users, providing "dial-tone." IXCs provide the long distance service between LEC companies.

Chapter 4

Cellular and PCS Mobile Radio

Introduction

Cellular and personal communication service (PCS) mobile radio system are wireless communication systems that provide for voice and data communication throughout a wide geographic area. The key to the success for cellular systems is their flexibility to increase the num-

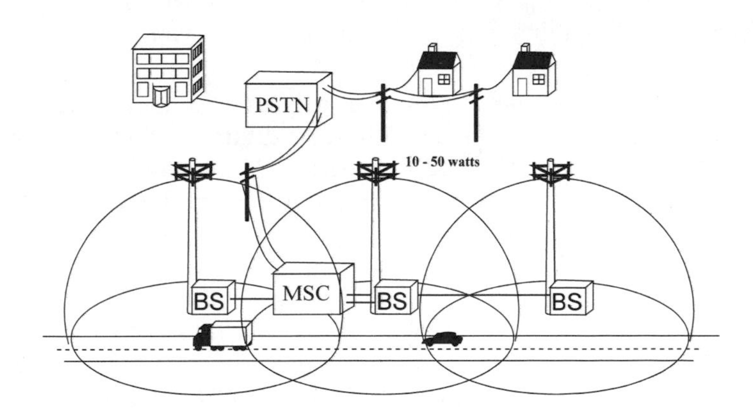

Figure 4.1 Cellular System

ber of channels to service the ever-increasing capacity demands of a growing customer base.

Figure 4.1 shows a basic cellular system. The cellular network connects mobile radios to each other or the public switched telephone network (PSTN) by using radio towers (base stations) that are connected to a mobile switching center (MSC). The mobile switching center can transfer calls to the PSTN.

When linked together to cover an entire metro area, the radio coverage areas (called cells) form a cellular structure resembling that of a honey comb. The cellular systems are designed to have overlap at each cell border to enable a "hand-off" from one cell to the next. As a customer (called a subscriber) moves through a cellular or a Personal Communications Service (PCS) system, the mobile switching center (MSC) coordinates and transfers calls from one cell to another and maintains call continuity.

The Difference between Cellular and PCS

Personal Communications Services (PCS) is a new category of frequencies at around 2 billion hertz (GHz) (actually 1900 MHz in the U.S.) that are used to provide services that can be offered by cellular systems. From a wireless industry perspective, PCS has been defined as a way of communicating! A way of communicating with people rather than phones. Your call is routed or re-routed to locate the person not a location. These systems are designed to screen your personal calls from your business calls, one-number dialing, messaging and other call features competitive with the wired world. As one marketing slogan said "Imagine what wireless can do for you!".

Marketing of any PCS service has one thing in common - digital. All of the new PCS systems take advantage of digital radio transmission capabilities to differentiate PCS from the original cellular. Ironically, almost all analog cellular systems are converting to digital radio technology to allow them to offer the same services that PCS systems can offer.

PCS and cellular and the perceived battle of "who's better" is exactly what the FCC set out to do - create competition! As the coverage increases for PCS and as the old cellular matures, the differences to attract new subscribers will certainly be personal!

Market Growth

Estimates indicate that over 28,000 new subscribers sign up for wireless service each day. The industry claims that it will be a $60 billion market by the year 2004.

Wireless voice growth has been astounding compared to the growth of other technologies. In 1984, cellular users numbered approximately 91,000. By the end of 1997 there were over 50 million users. This is remarkable when you consider that it took just 14 years for the industry to reach 50 million users, compared with television, which took 24 years to reach 50 million users, and radio, which took 39 years to reach the same level of usage. Perhaps even more remarkable is that it took over 75 years for landline telephone service to reach 50 million subscribers. While wireless calls account for only about 2% of all calls today, by 2002 that figure is expected to jump to 40% of all calls.

PCS deployment has fueled wireless growth in the last several years with most major cities adding a third and even a fourth wireless carrier. Increased competition from these carriers has brought overall rates down by 10%-15% in many markets. Handset pricing has been steadily declining as well. Mobile phones, installed in cars, were selling for over $3,000 in 1984. Today many carriers are offering free handheld units in exchange for service contracts. Added features and enhancements, including e-mail, faxing, pre-paid service and paging capabilities have all contributed to growth. Market penetration is between 14% - 24% in many of the larger markets. Interestingly, 90% of the population resides in the top 50 markets in the US.

Market segmentation has shifted from primarily business use to consumer usage. The average cellular bill is half of what it was in 1987, going from just over $100 to just over $47.

Subscribers	**1997**	**1998**	**1999**	**2000**	**2001**	**2002**
Cellular	50.6	61.5	67.9	73	77	93
PCS	3.2	7.5	12.7	18.5	24.9	32
TOTAL	**53.8**	**69**	**80.6**	**91.5**	**101.9**	**125**

In Millions

Figure 4.2., Cellular and PCS Market Growth (United States)
Source: Donaldson, Luftkin & Jenrette, Strategis Group, BIA, Insight Research, APDG Research

The outlook for continued wireless voice growth is extremely promising. Figure 4.2 depicts the expected growth of the industry.

Technologies

In early mobile radio telephone systems, one high-power transmitter served a large geographic area with a limited number of radio channels. Because each radio channel requires a certain frequency bandwidth (radio spectrum) and there is a very limited amount of radio spectrum available, this dramatically limits the number of radio channels which keeps the low serving capacity of such systems. For example, in 1976, New York City had only 12 radio channels to support 545 customers and a two-year long waiting list of typically 3,700 [1].

To conserve the limited amount of radio spectrum (maximum number of available radio channels), the cellular system concept was developed. Cellular systems allow reuse of the same channel frequencies many times within a geographic coverage area. The technique, called frequency reuse, makes it possible for a system to provide service to more customers (called system capacity) by reusing the channels that are available in a geographic area. In large systems such as the systems operating in New York City and Los Angeles, radio channel frequencies may be reused over 300 times. As systems start to become overloaded with many users, to increase capacity, the system can expand by simply adding more radio channels to the base station or by adding more cell cites with smaller coverage areas.

To minimize interference in this way, cellular system planners position the cell sites that use the same radio channel farthest away from each other. The distances between sites are initially planned by general RF signal propagation rules. But it is difficult to account for enough propagation factors to precisely position the towers, so the cell site position and power levels are usually adjusted later.

The acceptable distance between cells that use the same channels is determined by the distance to radius (D/R) ratio. The D/R ratio is the ratio of the distance (D) between cells using the same radio frequency to the radius (R) of the cells. In today's analog system, a typical D/R ratio is 4:6, a channel used in a cell with a one mile radius would not interfere with the same channel being reused at a cell 4.6 miles away. For some of the digital systems (such as CDMA or GSM), the reuse factor can be lower than 2.0.

Another technique, called cell splitting, helps to expand capacity gradually. Cells are split by adjusting the power level and/or using reduced antenna height to cover a reduced area. Reducing a coverage area by changing the RF boundaries of a cell site has the same effect as placing cells farther apart, and allows new cell sites to be added. However, the boundaries of a cell site vary with the terrain and land conditions, especially with seasonal variations in foliage. Coverage areas actually increase in fall and winter as the leaves fall from the trees.

When a cellular system is first established, it can effectively serve only a limited number of callers. When that limit is exceeded, callers experience system busy signals (known as blocking) and their calls cannot be completed. More callers can be served by adding more cells with smaller coverage areas - that is, by cell splitting. The increased number of smaller cells provides more available radio channels in a given area because it allows radio channels to be reused at closer geographical distances.

There are two basic types of systems: analog and digital. Analog systems commonly use FM modulation to transfer voice information and digital systems use some form of phase modulation to transfer digital voice and data information. Although analog systems are capable of providing many of the services that digital systems offer, digital systems offer added flexibility as many of the features can be created by software changes. The trend at the end of the 1990's was for analog systems to convert to digital systems.

To allow for the conversion from analog systems to digital systems, some cellular technologies allow for the use of dual mode or multimode mobile telephones. These handsets are capable of operating on an analog or digital radio channel, whichever is available. Most dual mode phones prefer to use digital radio channels, in the event both are available. This allows them to take advantage of the additional capacity and new features such as short messaging and digital voice quality, as well as offering greater capacity.

Cellular systems have several key differences that include the radio channel bandwidth, access technology type (FDMA, TDMA, CDMA), data signaling rates of their control channel(s) and power levels. Analog cellular systems have very narrow radio channels that vary from 10 kHz to 30 kHz. Digital systems channel bandwidth ranges from 30 kHz to 1.25 MHz. Access technologies determine how mobile telephones obtain service and how they share each radio channel. The data signaling rates determine how fast messages can be sent on control channels. The RF power level of mobile telephones and how the power level is controlled ordinarily determines how far away the mobile telephone can operate from the base station (radio tower).

Regardless of the size and type of radio channels, all cellular and PCS systems allow for full duplex operation. Full duplex operation is the ability to have simultaneous communications between the caller and the called person. This means a mobile telephone must be capable of simultaneously transmitting and receiving to the radio tower. The radio channel from the mobile telephone to the radio tower is called the uplink and the radio transmission channel from the base station to the mobile telephone is called the downlink. The uplink and downlink radio channels are normally separated by 45 MHz to 80 MHz.

One of the key characteristics of cellular systems is their ability to handoff (also called handover) calls from one radio tower to another while a call is in process. Handoff is a automatic process that is a result of system monitoring and short control messages that are sent between the mobile phone and the system while the call is in progress. The control messages are so short that the customer usually cannot perceive that the handoff has occurred.

Analog Cellular

Analog cellular systems are regularly characterized by their use of analog modulation (commonly FM modulation) to transfer voice information. Ironically, almost all analog cellular systems use separate radio channels for sending system control messages. These are digital radio channels.

In early mobile radio systems, a mobile telephone scanned the limited number of available channels until it found an unused one, which allowed it to initiate a call. Because the analog cellular systems in use today have hundreds of radio channels, a mobile telephone cannot scan them all in a reasonable amount of time. To quickly direct a mobile telephone to an available channel, some of the available radio channels are dedicated as control channels. Most cellular systems use two types of radio channels, control channels and voice channels. Control channels carry only digital messages and signals, which allow the mobile telephone to retrieve system control information and compete for access.

Control channels only carry control information such as paging (alert) and channel assignment messages. Voice channels are primarily used to transfer voice information. However, voice channels must also be capable of sending and receive some digital control messages to allow for necessary frequency and power changes during a call.

Current analog systems serve only one subscriber at a time on a radio channel, so system capacity is influenced by the number of radio channels available. However, a typical subscriber uses the system for only a few minutes a day, so on a daily basis, many subscribers share a single channel. As a rule, 20 - 32 subscribers share each radio channel [2], depending upon the average talk time per hour per subscriber. Generally, a cell with 50 channels can support 1000 - 1600 subscribers.

The basic operation of an analog cellular system involves initiation of the phone when it is powered on, listening for paging messages (idle), attempting access when required and conversation (or data) mode.

When a mobile telephone is first powered on, it initializes itself by searching (scanning) a predetermined set of control channels and then tuning to the strongest one. During the initialization mode, it listens to messages on the control channel to retrieve system identification and setup information.

After initialization, the mobile telephone enters the idle mode and waits to be paged for an incoming call and senses if the user has initiated (dialed) a call (access). When a call begins to be received or initiated, the mobile telephone enters system access mode to try to access the system via a control channel. When it gains access, the control channel sends an initial voice channel designation message indicating an open voice channel. The mobile telephone then tunes to the designated voice channel and enters the conversation mode. As the mobile telephone operates on a voice channel, the system uses Frequency Modulation (FM) similar to commercial broadcast FM radio. To send control messages on the voice channel, the voice information is either replaced by a short burst (blank and burst) message or in some systems, control messages can be sent along with the audio signal.

A mobile telephone's attempt to obtain service from a cellular system is referred to as "access". Mobile telephones compete on the control channel to obtain access from a cellular system. Access is attempted when a command is received by the mobile telephone indicating the system needs to service that mobile telephone (such as a paging message indicating a call to be received) or as a result of a request from the user to place a call. The mobile telephone gains access by monitoring the busy/idle status of the control channel both before and during transmission of the access attempt message. If the channel is available, the mobile station begins to transmit and the base station simultaneously monitors the channel's busy status. Transmissions must begin within a prescribed time limit after the mobile station finds that the control channel access is free, or the access attempt is stopped on the assumption that another mobile telephone has possibly gained the attention of the base station control channel receiver.

If the access attempt succeeds, the system sends out a channel assignment message commanding the mobile telephone to tune to a cellular voice channel. When a subscriber dials the mobile telephone to initiate a call, it is called "origination". A call origination access attempt message is sent to the cellular system that contains the dialed digits, identity information along with other information. If the system allows service, the system will assign a voice channel by sending a voice channel designator message, if a voice channel is available. If the access attempt fails, the mobile telephone waits a random amount of time before trying again. The mobile station uses a random number generating algorithm internally to determine the random time to wait. The design of the system minimizes the chance of repeated collisions between different mobile stations which are both trying to access the control channel, since each one waits a different random time interval before trying again if they have already collided on their first, simultaneous attempt.

To receive calls, a mobile telephone is notified of an incoming call by a process called paging. A page is a control channel message which contains the telephone's Mobile Identification Number (MIN) or telephone number of the desired mobile phone. When the telephone determines it has been paged, it responds automatically with a system access message that indicates its access attempt is the result of a page message and the mobile telephone begins to ring to alert the customer of an incoming telephone call. When the customer answers the call (user presses "SEND" or "TALK"), the mobile telephone transmits a service request to the system to answer the call. It does this by sending the telephone number and an electronic serial number to provide the users identity.

After a mobile telephone has been commanded to tune to a radio voice channel, it sends mostly voice or other customer information. Periodically, control messages may be sent between the base station and the mobile telephone. Control messages may command the mobile telephone to adjust it's power level, change frequencies, or request a special service (such as three way calling).

To conserve battery life, a mobile phone may be permitted by the base station to only transmit when it senses the mobile telephone's user is talking. When there is silence, the mobile telephone may stop transmitting for brief periods of time (several seconds). When the mobile telephone user begins to talk again, the transmitter is turned on again. This is called discontinuous reception.

Figure 4.3 shows a basic analog cellular system. This diagram shows that there are two types of radio channels: control channels and voice channels. Control channels ordinarily use frequency shift keying (FSK) to send control messages (data) between the mobile phone and the base station. Voice channels normally use FM modulation with brief bursts of digital information to allow control messages (such as handoff) during conversation. Base stations usually have two antennas for receiving and one for transmitting. Dual receiver antennas increases the ability to receive the radio signal from mobile telephones which usually have a much lower transmitter power level than the transmitters in the base station. Base stations are connected to a mobile switching center (MSC) commonly by a high speed telephone line or microwave radio system. This interconnection must allow both voice and control information to be exchanged between the switching system and the base station. The MSC is connected to the telephone network to allow mobile telephones to be connected to standard landline telephones.

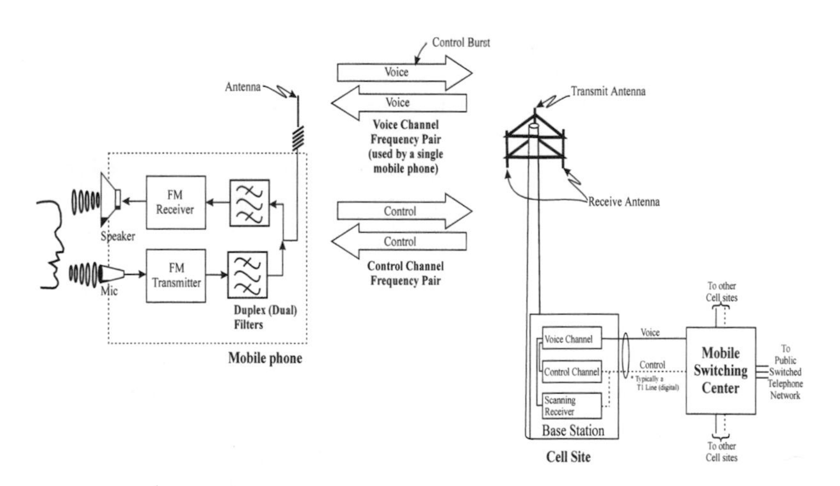

Figure 4.3 Analog Cellular Technology

Digital Cellular

Digital cellular systems are often characterized by their type of access technology (TDMA or CDMA). The access technology determines how that digital information is transferred to and from the cellular system.

Digital cellular systems can ordinarily serve several subscribers on a single radio channel at the same time. Depending on the type of system, this can range from 3 to over 20. To allow this, almost all digital cellular systems share the fundamental characteristics of digitizing and compressing voice information to accomplish this. This allows a single radio channel to be divided into several sub-channels (communication channels). Each communication channel can serve a single customer.

Because each subscriber typically uses the cellular system for only a few minutes a day, several subscribers can share each one of these communication channels during the day. As a rule, 20 - 32 subscribers can share each communication channel; so if a digital radio channel has eight communications channels (sub-channels), a cell site with 25 radio channels can support 4000 to 6400 subscribers.

Digital cellular systems use two key types of communication channels, control channels and voice channels. A control channel on a digital system is usually one of the sub-channels on the radio channel. This allows digital systems to combine a control channel and one or more voice channels on a single radio channel. The portion of the radio channel that is dedicated as a control channel carries only digital messages and signals which allow the mobile telephone to retrieve system control information and compete for access. The other sub-channels on the radio channel carry voice or data information.

The basic operation of a digital cellular system involves initiation of the phone when it is powered on, listening for paging messages (idle), attempting access when required and conversation (or data) mode.

When a digital mobile telephone is first powered on, it initializes itself by searching (scanning) a predetermined set of control channels and then tuning to the strongest one. During the initialization mode, it listens to messages on the control channel to retrieve system identification and setup information. Compared to analog systems, digital systems have more communication channels. Digital systems have more communication channels, where control channels may be located, than analog systems. This can result in the mobile phone taking more time to search for control channels. To quickly direct a mobile telephone to an available control channel, digital systems use several processes to help a mobile telephone to find an available control channel. These include having the phone memorize its last successful control channel location, a table of likely control channel locations and a mechanism for pointing to the location of a control channel on any of the operating channels.

After a digital mobile telephone has initialized, it enters an idle mode where it waits to be paged for an incoming call or for the user to initiate a call. When a call begins to be received or initiated, the mobile telephone enters system access mode to try to access the system via a control channel. When it gains access, the control channel sends a digital traffic channel designation message indicating an open communications channel. This channel may be on a different time slot on the same frequency or to a time slot on a different frequency. The digital mobile telephone then tunes to the designated communications channel and enters the conversation mode. As the mobile telephone operates on a digital voice channel, the digital system commonly uses some form of phase modulation (PM) to send and receive digital information.

A mobile telephone's attempt to obtain service from a cellular system is referred to as "access". Digital mobile telephones compete on the control channel to obtain access from a cellular system. Access is attempted when a command is received by the mobile telephone indicating the system needs to service that mobile telephone (such as a paging message indicating a call to be received) or as a result of a request from the user to place a call. Digital mobile telephones usually have the ability to validate their identities more securely during access than analog mobile telephones. This is made possible by a process called authentication. Authentication processes and share secret data between the digital mobile phone and the cellular system.

If the authentication is successful, the system sends out a channel assignment message commanding the mobile telephone to change to a new communication channel and conversation can begin.

After a mobile telephone has been commanded to tune to a radio voice channel, it sends digitized voice or other customer data. Periodically, control messages may be sent between the base station and the mobile telephone. Control messages may command the mobile telephone to adjust its power level, change frequencies, or request a special service (such as three way calling). To send control messages while the digital mobile phone is transferring digital voice, the voice information is either replaced by a short burst (called blank and burst or fast signaling), or else control messages can be sent along with the digitized voice signals (called slow signaling).

Most digital telephones automatically conserve battery life as they transmit only for short periods of time (bursts). In addition to savings through digital burst transmission, digital phones ordinarily have the capability of discontinuous transmission that allows the inhibiting of the transmitter during periods of user silence. When the mobile telephone user begins to talk again, the transmitter is turned on again. The combination of the power savings allows some digital mobile telephones to have 2 to 5 times the battery life in the transmit mode.

Digital technology increases system efficiency by voice digitization, speech compression (coding), channel coding, and the use of spectrally efficient radio signal modulation.

Voice digitization produces a data rate of approximately 64 kilobits per second (kbp/s). Because transmitting a digital signal via radio requires about 1 Hz of radio bandwidth for each kbp/s, an uncompressed digital voice signal would require more than 64 kHz of radio bandwidth. Without compression, this bandwidth would make digital transmission less efficient than analog FM cellular, which uses only 25-30 kHz for a single voice channel. Therefore, digital systems compress speech information using a speech coder (Coder/Decoder). Speech coding removes redundancy in the digital signal and attempts to ignore data patterns that are not characteristic of the human voice. The result is a digital signal which represents the voice audio frequency spectrum content, not a waveform.

A speech coder characterizes the input signal. It looks up codes in a code book table which represents various digital patterns to chose the pattern which comes closest to the input digitized signal. The amount of digitized speech compression used in digital cellular systems varies. For the IS-136 TDMA system, the compression is 8:1. For CDMA, the compression varies from 8:1 to 64:1 depending on speech activity. GSM systems compress the voice by 5:1.

As a general rule, with the same amount of speech coding analysis, the fewer bits used to characterize the waveform, the poorer the speech quality. If the complexity (signal processing) of the speech coder can be increased, it is possible to get improved voice quality with fewer bits.

Voice digitization and speech coding take processing time. Typically, speech frames are digitized every 20 msec and inputted to the speech coder. The compression process, time alignment with the radio channel, and decompression at the receiving end all delay the voice signal. The combined delay can add up to 50-100 msec. Although such a delay is not usually noticeable in two-way conversation, it can cause an annoying echo when a speakerphone is used, or the side tone of the signal is high (so the users can hear themselves). However, an echo canceller can be used in the MSC to process the signal and remove the echo.

Once the digital speech information is compressed, control information bits must be added along with extra bits to protect from errors that will be introduced during radio transmission. The combined digital signal (compressed digitized voice and control information) is sent to the radio modulator where it is converted to a digitized RF signal. The efficient conversion to the RF signal constantly involves some form of phase shift modulation.

Figure 4.4 shows a basic digital cellular system. This diagram shows that there usually is only one type of digital radio channel called a digital traffic channel (DTC). The digital radio channel is ordinarily subdivided into control channels and digital voice channels. Both the control channels and voice channels use the same type of digital modulation to send control and content data between the mobile phone and the base station. When used for voice, the digital signal is usually a compressed digital signal that is from a speech coder. When conversation is in progress, some of the digital bits are usually dedicated for control information (such as handoff). Similar to analog systems, digital base stations have two antennas to increases the ability to receive weak radio signals from mobile telephones. Base stations are connected to a mobile switching center (MSC) normally by a high speed telephone line or microwave radio system. This interconnection may allow compressed digital information (directly from the speech coder) to increase the number of voice channels that can be shared on a single connection line. The MSC is connected to the telephone network to allow mobile telephones to be connected to standard landline telephones.

Commercial Systems

There are many types of analog and digital cellular systems in use throughout the world. Analog systems systems include AMPS, TACS, JTACS, NMT, MCS and CNET. Digital systems include GSM, IS-136 TDMA and CDMA.

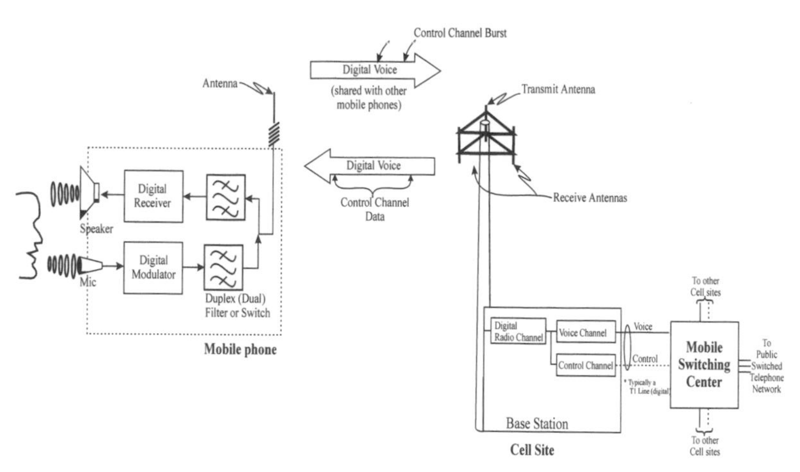

Figure 4.4 Digital Cellular Technology

Advanced Mobile Phone Service (AMPS)

Advanced mobile phone service (AMPS) was the original analog cellular system in the United States. It is still in widespread use and in 1997, AMPS systems were operating in over 72 countries [3]. The AMPS system continues to evolve to allow advanced features such as increased standby time, narrowband radio channels, and anti-fraud authentication procedures.

In 1974, 40 MHz of spectrum was allocated for cellular service [4] which provided only 666 channels. In 1986, an additional 10 MHz of spectrum was added to facilitate expansion [5] of the system to 832 channels.

The frequency bands for the AMPS system are 824 MHz to 849 MHz (uplink) and 869 MHz to 894 MHz (downlink). Of the 832 channels, AMPS systems are divided into A and B bands to allow for two different service providers. There are two types of radio channels in an AMPS system; dedicated control channels and voice channels. On each system (A or B), mobile telephones scan and tune to one of 21 dedicated control channels to listen for pages and compete for access to the system. The control channel continuously sends system identification information and access control information. Although the control channel data rate is 10 kbps, messages are repeated 5 times which reduces the effective channel rate to below 2 kbps. This allows a control channel to send 10 to 20 pages per second.

The AMPS cellular system is frequency duplex with its channels separated by 45 MHz. The control channel and voice channel signaling is transferred at 10 kbps. AMPS cellular phones have three classes of maximum output power. A class 1 mobile telephone has a maximum power output of 6 dBW (3 Watts), class 2 has a maximum output power of 2 dBW (1.6 Watts), and the class 3 units are capable of supplying only -2 dBW (.6 Watts). The output power can be adjusted in 4 dB steps and has a minimum output power of -22 dBW (approximately 6 milliwatts).

Total Access Communication System (TACS)

The total access communication system (TACS) is very similar to the US EIA-553 AMPS system. Its primary differences include changes to the radio channel frequencies, radio channel bandwidths, and data signaling rates. The TACS was introduced to the U.K. in 1985. Since its introduction in the UK in 1985, over 25 countries offer TACS service. The introduction of the TACS system was very successful and the system was expanded to add more channels through what is called Extended TACS (ETACS).

The TACS system was created by enhancing the efficiency of the AMPS cellular system radio channels. These enhancements include allowing a smaller radio channel bandwidth, reduced data speed for the signaling channel, and adding new features.

The frequency ranges of most TACS systems are 890 MHz to 915 MHz for the uplink and 935 MHz to 960 MHz for the downlink. The TACS system was initially allocated 25 MHz although 10 MHz of the 25 MHz was reserved for future pan-European systems in the UK. An additional 16 MHz of radio channel bandwidth was added to allow for Extended TACS (ETACS). The ETACS system is a frequency duplex system with its channels separated by 45 MHz.

The control channel and voice channel signaling is transferred at 8 kbps. There are four power classes for ETACS mobile telephones. Class 1 mobile telephones have a maximum output of 10 Watts, class 2 has 3 Watts, class 3 has 1.2 Watts, and class 4 has .6 Watts. Similar to AMPS, mobile telephones can be adjusted in 4 dB steps and have a minimum transmit power level of approximately 6 milliwatts.

The TACS system has also been modified for use in Japan. This Japanese version is called JTACS. The only significant changes were the frequency bands and number of channels. The TACS system has also been modified to create the Narrowband TACS (NTACS) system. NTACS reduced the radio channel bandwidth from 25 kHz to 12.5 kHz and changed the in-band 8 kbps signaling on the voice channel to 100 bps sub-band digital signaling.

Nordic Mobile Telephone (NMT)

There are two Nordic Mobile Telephone (NMT) systems; NMT 450 which is a low capacity system and NMT 900 which is a high capacity system. The Nordic mobile telephone (NMT) system was developed by the telecommunications administrations of Sweden, Norway, Finland, and Denmark to create a compatible mobile telephone system in the Nordic countries [6]. The first commercial NMT 450 cellular system was available at the end of 1981. Due to the rapid success of the initial NMT 450 system and limited capacity of the original system design, the NMT 900 system version was introduced in 1986. There are now over 40 countries that have NMT service available. Some of these countries use different frequency bands or reduced number of channels.

The NMT 450 system uses a lower frequency (450 MHz) and higher maximum transmitter power level which allows a larger cell site coverage areas while the NMT 900 system uses a higher frequency (approximately the same 900 MHz band used for TACS and GSM) and a lower maximum transmitter power which increases system capacity. NMT 450 and NMT 900 systems can co-exist which permits them to use the same switching center [7]. This allows some NMT service providers to start offering service with an NMT 450 system and progress up to a NMT 900 system when the need arises.

Some operations of the NMT systems are very different from most other cellular systems. When NMT mobile telephones access the cellular system, they can either find an unused voice channel and negotiate access directly or begin conversation without the assistance of a dedicated control channel. Because scanning for free voice channels can be very time consuming, the NMT 900 system does allow for the use of a dedicated control channel called the calling channel. The NMT 900 system also allows discontinuous reception, which increases the standby time of the portable phones.

The NMT 450 system is frequency duplex with 180 channels (except Finland which only has 160 channels) [8]. The radio channel bandwidth is 25 kHz and the frequency duplex spacing is 10 MHz. The NMT 900 system has 1000 channels for 999 interleaved channels.

Signaling on the NMT systems is performed at 1200 bps on the control (calling) channel (NMT 900) and voice channel. Because of the slow signaling rate and robust error detection/correction capability, no repeated messages are necessary.

NMT 450 base stations can transmit up to 50W. This high power combined with the lower 450 MHz frequency allows cell site size of up to approximately 40 km radius. NMT 900 base stations are limited to a maximum of 25W which allows a maximum cell size radius of up to approximately 20 km [9].

There are three power levels (high, medium, and low) for NMT mobile phones and two power levels (high and low) for portables. NMT 450 mobile telephone power levels are: High 15W, Medium 1.5W, Low 0.15W. NMT 450 portable telephones; High 1.0W, Low .1W. NMT 900 mobile telephones: High 6.0W, Medium 1.0W, Low .1W and NMT 900 portable telephones: High 1.0W, Low 0.1W.

The NMT system is unique as it has various types of anti-fraud protection. NMT mobile telephones hold a three digit password which is stored in the telephone and cellular switching center and is unknown to the customer. This password is sent to the cellular system during system access along with the mobile telephone number. The NMT system has also added a Subscriber Identity Security (SIS) system that provides additional anti-fraud protection. Not all NMT telephones have SIS capability.

Narrowband AMPS (NAMPS)

Narrowband Advanced Mobile Phone Service (NAMPS) is an analog cellular system which was commercially introduced by Motorola in late 1991 and is currently being deployed worldwide. Like the existing AMPS technology, NAMPS uses analog FM radio for voice transmissions. The distinguishing feature of NAMPS is its use of a "narrow" 10 kHz bandwidth for radio channels, a third of the size of AMPS channels. Because more of these narrower radio channels can be installed in each cell site, NAMPS systems can serve more subscribers than

AMPS systems without adding new cell sites. NAMPS also shifts some control commands to the sub-audible frequency range to facilitate simultaneous voice and data transmissions.

In 1991, the first NAMPS standard, named IS-88, evolved from the US AMPS specification (EIA-553). The IS-88 standard identified parameters needed to begin designing NAMPS radios, such as radio channel bandwidth, type of modulation, and message format. During development, the NAMPS specification benefited from the narrowband JTACS radio system specifications. During the following years, advanced features such as ESN authentication, caller ID, and short messaging were added to the NAMPS specification.

Japanese Mobile Cellular System (MCS)

Japan launched the world's first commercial cellular system in 1979. Because this system had achieved great success, several different types of cellular systems have evolved in Japan. These include the MCS-L1, MCS-L2, JTACS and NTACS systems.

The MCS-L1 was the first cellular system in Japan which was developed and operated by NTT. The system operates in the 800 MHz band. The channel bandwidth is 25 kHz and the signaling is at 300 bps. The control channels are simulcast from all base stations in the local area. This limits the maximum capacity of the MCS-L1 system.

Because the MCS-L1 system could only serve a limited number of customers, the MCS-L2 system was developed. It uses the same frequency bands as the MCS-L1 system. The radio channel bandwidth was reduced from 25 kHz to 12.5 kHz with 6.25 kHz interleaving. This gives the MCS-L2 system 2,400 channels. The control channels transfer information at 2,400 bps and the voice channels can use either in-band (blank and burst) signaling at 2,400 bps or sub-band digital audio signaling at 150 bps. MCS-L2 mobile telephones have diversity reception (similar to diversity receive used in base stations). While this increases the cost and size of the mobile telephones, it also increases the performance and range of the cellular system.

CNET

CNET is an analog cellular system that is used in Germany, Portugal, and South Africa [10]. The first CNET system started operation in

Germany in 1985. The primary objective of the CNET system was to bridge the gap of cellular systems in Germany until the digital European system could be introduced [11].

The CNET system operates at 450 MHz with 4.44 MHz transmit and receive bands. The frequency bands are 461.3 to 465.74 MHz and 451.3 to 455.74 MHz. The primary channel bandwidth is 20 kHz with 10 kHz channel interleaving.

The CNET system continuously exchanges digital information between the mobile telephone and the base station. Every 12.5 msec, 4 bits of information are sent during compressed speech periods [12]. CNET mobile telephones also use an Identification Card (IC) which slides into the telephone to identify the customer. This allows customers to use any compatible CNET telephone.

MATS-E

The MATS-E system is used in France and Kuwait [13]. The MATS-E system combines many of the features used in different cellular systems. MATS-E uses the standard European mobile telephone frequency bands; 890-915 MHz and 935-960 MHz. The channel bandwidth is 25 kHz which provides 1,000 channels. The MATS-E is a frequency duplex system separated by 45 MHz. Each cell site has at least one dedicated control channel with a signaling rate of 2400 bps. Voice channels use FM modulation with sub-band digital audio signaling with a data rate of 150 bps

Global System for Mobile Communication (GSM)

The Global System for Mobile Communications (GSM) system is a global digital radio system which uses Time Division Multiple Access (TDMA) technology. GSM is a digital cellular technology which was initially created to provide a single-standard pan-European cellular system. GSM began development in 1982, and the first commercial GSM digital cellular system was activated in 1991. GSM technology has evolved to be used in a variety of systems and frequencies (900 MHz, 1800 MHz and 1900 MHz) including Personal Communications Services (PCS) in North America and Personal Communications Network (PCN) systems throughout the world. In 1998, over 140 countries offered GSM service.

The GSM system is a digital-only system and was not designed to be backward-compatible with the established analog systems. The GSM radio band is shared temporarily with analog cellular systems in some European nations.

When communicating in a GSM system, users can operate on the same radio channel simultaneously by sharing time slots. The GSM cellular system allows eight mobile telephones to share a single 200 kHz bandwidth radio carrier waveform for voice or data communications. To allow duplex operation, GSM voice communication is conducted on two 200 kHz wide carrier frequency waveforms.

The GSM system has several types of control channels that carry system and paging information, and coordinates access like the control channels on analog systems. The GSM digital control channels have many more capabilities than analog control channels such as broadcast message paging, extended sleep mode, and others. Because the GSM control channels use only a portion (one or more slots), they typically co-exist on a single radio channel with other time slots that are used for voice communication.

Voice channels can be either full rate or half rate. Full rate GSM systems assign one time slot per frame to each user , allowing eight users to simultaneously share a radio channel. GSM is designed so that it can easily accommodate a future half-rate speech coder (a digital speech coder which produces good speech quality at half of the bit rate of the present speech coder) which is expected to emerge from the research laboratory in the next few years. Half rate GSM systems assign one time slot every other frame to allow up to 16 users to share a radio channel.

A GSM carrier transmits at a bit rate of 270 kbps, but a single GSM digital radio channel or time slot is capable of transferring only 1/8 th of that, about 33 kbps of information (actually less than that, due to the use of some bit time for non-information purposes such as synchronization bits).

Time intervals on full rate GSM channels are divided into frames with eight time slots on two different radio frequencies. One frequency is for transmitting from the mobile telephone; the other is for receiving to the mobile telephone. During a voice conversation at the mobile set, one time slot is dedicated for transmitting, one for receiving, and six remain idle. The mobile telephone uses the idle time slots to measure the signal strength of surrounding cell carrier frequencies.

On the 900 MHz band, GSM digital radio channels transmit on one frequency and receive on another frequency 45 MHz higher, but not at the same time. On the 1.9 GHz band, the difference between transmit and receive frequencies is 80 MHz. The mobile telephone transmits a burst of data on one frequency, then receives a burst on another frequency, and is briefly idle before repeating the process.

North American TDMA (IS-136 TDMA)

The North American TDMA system (IS-136) is a digital system that uses TDMA access technology. It evolved from the IS-54 specification that was developed in North America in the late 1980's to allow the gradual evolution of the AMPS system to digital service. The IS-136 system is sometimes referred to as Digital AMPS (DAMPS) or North American digital cellular (NADC).

In 1988, the Cellular Telecommunications Industry Association created a development guideline for the next generation of cellular technology for North America. This guideline was called the User Performance Requirements (UPR) and the Telecommunications Industry Association (TIA) used this guideline to create a TDMA digital standard, called IS-54. This digital specification evolved from the original EIA-553 AMPS specification. The first revision of the IS-54 specification (Rev 0) identified the basic parameters (e.g. time slot structure, type of radio channel modulation, and message formats) needed to begin designing TDMA cellular equipment. There have been several enhancements to IS-54 since its introduction and in 1995, IS-54 was incorporated as part of the IS-136 specification.

A primary feature of the IS-136 systems is their ease of adaptation to the existing AMPS system. Much of this adaptability is due to the fact that IS-136 radio channels retain the same 30 kHz bandwidth as AMPS system channels. Most base stations can therefore replace TDMA radio units in locations previously occupied by AMPS radio units. Another factor in favor of adaptability is that new dual mode mobile telephones were developed to operate on either IS-136 digital traffic (voice and data) channels or the existing AMPS radio channels as requested in the CTIA UPR document. This allows a single mobile telephone to operate on any AMPS system and use the IS-136 system whenever it is available.

The IS-136 specification concentrates on features that were not present in the earlier IS-54 TDMA system. These include longer standby time, short message service functions, and support for small private or residential systems that can coexist with the public systems. In addition, IS-136 defines a digital control channel to accompany the Digital Traffic Channel (DTC). The digital control channel allows a mobile telephone to operate in a single digital-only mode. Revision A of the IS-136 specification now supports operation in the 800Mhz range for the existing AMPS and DAMPS systems as well as the newly allocated 1900MHz bands for PCS systems. This permits dual band, dual mode phones (800 MHz and 1900 MHz for AMPS and DAMPS). The primary difference between the two bands is that mobile telephones cannot transmit using analog signals at 1900MHz.

The IS-136 cellular system allows for mobile telephones to use either 30 kHz analog (AMPS) or 30 kHz digital (TDMA) radio channels. The IS-136 TDMA radio channel allows multiple mobile telephones to share the same radio frequency channel by time sharing. All IS-136 TDMA digital radio channels are divided into frames with six time slots. The time slots used for the correspondingly numbered forward and reverse channels are time-related so that the mobile telephone does not simultaneously transmit and receive.

Similar to the GSM system, voice channels can have differing rates; full rate or half rate. Full rate IS-136 systems allow three simultaneous users per radio channel and half rate systems allow six users to share a radio channel.

The IS-136 system allows a standard time slot on a TDMA radio channel to be used as a digital control channel (DCC). DCC carries the same system and paging information as the analog control channel (ACC). In addition to the control messages, DCC has more capabilities than the ACC such as extended sleep mode, short message service (SMS), private and public control channels, and others.

The total bit rate of the carrier frequency waveform is 48.6 kbps. This time sharing and the use of some bits for synchronization and low bit-rate data transmission results in a user-available data rate of 13 kbps. Some of the 13 kbps are used for error detection and correction, so only 8 kbps of data are available for full rate digitally coded speech.

The RF power levels for the mobile phones are almost exactly the same as for the AMPS telephones. The primary difference in the power levels is a reduction in minimum power level that mobile telephones can be instructed to reduce to. This allows for very small cell coverage areas, typically the size of cells that would be used for wireless office or home cordless systems.

Extended TDMA (E-TDMA)™

Extended TDMA™ was developed by Hughes Network Systems in 1990 as an extension to the existing IS-136 TDMA industry standard. ETDMA uses the existing TDMA radio channel bandwidth and channel structure and its receivers are tri-mode as they can operate in AMPS, TDMA, or ETDMA modes. While a TDMA system assigns a mobile telephone fixed time slot numbers for each call, ETDMA dynamically assigned time slots on an as needed basis. The ETDMA system contains a half-rate speech coder (4 kb/s) that reduces the number of information bits that must be transmitted and received each second. This makes use of voice silence periods to inhibit slot transmission so other users may share the transmit slot. The overall benefit is that more users can share the same radio channel equipment and improved radio communications performance. The combination of a low bit rate speech coder, voice activity detection, and interference averaging increases the radio channel efficiency to beyond 10 times the existing AMPS capacity.

ETDMA radio channels are structured into the same frames and slots structures as the standard IS-54 radio channels. Some or all of the time slots on all of the radio channels are shared for ETDMA communication, which is similar to IS-54 and IS-136 radio channels, or else slots can be shared on different frequencies. When a Mobile telephone is operating in extended mode, the ETDMA system must continually coordinate time slot and frequency channel assignments. The ETDMA system performs this by using a time slot control system. On an ETDMA capable radio channel some of the time slots are dedicated as control slots on an as needed basis. ETDMA systems can assign either an AMPS channel, a TDMA full-rate or half-rate channel, or an ETDMA channel. The existing 30 kHz AMPS control channels are used to assign analog voice and digital traffic channels

In an ETDMA system, some of the radio channels include a control slot which coordinates time slot allocation. This usually accounts for an estimated 15% of available time slots in a system. The control time slots assign an ETDMA subscriber to voice time slots on multiple radio channels.

ETDMA uses the following process to allocate time slots from moment to moment as needed. The cellular radio maintains constant communications with the Base Station through the control time slot. When a conversation begins, the cellular radio uses the control slot to request a voice time slot from the Base Station. Through the control slot, the Base Station assigns a voice time slot and sets the cellular radio to transmit in that assigned voice time slot. During each momentary lull in phone conversation, the transmitting cellular radio gives up its voice time slot, which is then placed back into the Base Station's pool of available time slots.

When a cellular radio is ready to receive a voice conversation, the Base Station uses the control slot to tell it which voice time slot has the conversation being sent. The cellular radio receiver then tunes to the appropriate slot. Through the control slot, the Base Station constantly monitors the cellular radio to determine whether it has given up a slot or needs a slot. In turn, the cellular radio constantly monitors the control slot to learn which time slot contains voice conversation being sent to it.

Code Division Multiple Access (IS-95 CDMA)

Code Division Multiple Access (CDMA) system (IS-95) is a digital cellular system that uses CDMA access technology. IS-95 technology was initially developed by Qualcomm in the late 1980's. CDMA cellular service began testing in the United States in San Diego, California during 1991. In 1995, IS-95 CDMA commercial service began in Hong Kong and now several CDMA systems are operating throughout the world, including a 1.9 GHz all-digital system in the USA which has been operating since November, 1996.

Spread spectrum radio technology has been used for many years in military applications. CDMA is a particular form of spread spectrum radio technology. In 1989, CDMA spread spectrum technology was presented to the industry standards committee but it did not meet with immediate approval. The standards committee had just resolved a two-year debate between TDMA and FDMA and was not eager to consider another access technology.

The IS-95 CDMA system allows for voice or data communications on either a 30 kHz AMPS radio channel (when used on the 800 MHz cellular band) or a new 1.23 MHz CDMA radio channel. The IS-95

CDMA radio channel allows multiple mobile telephones to communicate on the same frequency at the same time by special coding of their radio signals.

CDMA radio channels carry control, voice, and data signals simultaneously by dividing a single traffic channel (TCH) into different subchannels. Each of these channels is identified by a unique code. When operating on a CDMA radio channel, each user is assigned to a code for transmission and reception. Some codes in the TCH transfer control channel information, and some transfer voice channel information.

The control channel which is part of a digital traffic channel on a CDMA system has new advanced features. This digital control channel (DCC) carries system and paging information, and coordinates access similar to the analog control channel (ACC). The DCC has many more capabilities than the ACC such as a precision synchronization signal, extended sleep mode, and others. Because each CDMA radio channel has many codes, more than one control channel can exist on a single CDMA radio channel and the CDMA control channels co-exist with other coded channels that are used for voice.

The IS-95 CDMA cellular system has several key attributes that are different from other cellular systems. The same CDMA radio carrier frequencies may be optionally used in adjacent cell sites, which eliminates the need for frequency planning, the wide-band radio channel provides less severe fading, which the inventors claim results in consistent quality voice transmission under varying radio signal conditions. The CDMA system is compatible with the established access technology, and it allows analog (EIA-553) and dual mode (IS-95) subscribers to use the same analog control channels. Some of the voice channels are converted to CDMA digital transmissions, allowing several users to be multiplexed (shared) on a single RF channel. As with other digital technologies, CDMA produces capacity expansion by allowing multiple users to share a single digital RF channel.

The IS-95 CDMA radio channel divides the radio spectrum into wide 1.23 MHz digital radio channels. CDMA radio channels differ from those of other technologies in that CDMA multiplies (and therefore spreads the spectrum bandwidth of) each signal with a unique pseudo-random noise (PN) code that identifies each user within a radio channel. CDMA transmits digitized voice and control signals on the same frequency band. Each CDMA radio channel contains the signals of many ongoing calls (voice channels) together with pilot, synchronization, paging, and access (control) channels. Digital mobile telephones select the signal they are receiving by correlating (matching) the received signal with the proper PN sequence. The correlation enhances the power level of the selected signal and leaves others unenhanced.

Each IS-95 CDMA radio channel is divided into 64 separate logical (PN coded) channels. A few of these channels are used for control, and the remainder carry voice information and data. Because CDMA transmits digital information combined with unique codes, each logical channel can transfer data at different rates (e.g. 4800 b/s, 9600 b/s).

CDMA systems use a maximum of 64 coded (logical) traffic channels, but they cannot always use all of these. A CDMA radio channel of 64 traffic channels can transmit at a maximum information throughput rate of approximately 192 kbps [14], so the combined data throughput for all users cannot exceed 192 kbps. To obtain a maximum of 64 communication channels for each CDMA radio channel, the average data rate for each user should approximate 3 kbps. If the average data rate is higher, less than 64 traffic channels can be used. CDMA systems can vary the data rate for each user dependent on voice activity (variable rate speech coding), thereby decreasing the average number of bits per user to about 3.8 kbps [15]). Varying the data rate according to user requirement allows more users to share the radio channel, but with slightly reduced voice quality. This is called soft capacity limit.

Services

There are three basic services offered by cellular systems; voice, messaging and data. Advanced services such as voice mail and paging are often bundled into a basic service program.

Voice

The most well known application for wireless communications is voice communications. Voice communication can be telephony; wide area (cellular), business location (wireless office) or home cordless (residential) or voice paging, dispatch (fleet coordination) or group voice (audio broadcasting). Service rates for voice applications typically involve an initial connection charge, basic monthly minimum fee,

Activation Fee	$40.00
Monthly Access Fee	$25.00
Airtime Charges	$0.25 per minute
Roaming Charges	$0.60 per minute
Call Waiting, Call Forwarding, 3-Way	Free

Figure 4.5 Typical Cellular Service Rate Plan

more likely a monthly access fee that includes some free airtime minutes, plus an airtime usage charge. When the customer uses service in a system other than their home registered system (roaming), there may be a daily roaming fee and/or a higher per minute roaming usage fee. A sample service charge is shown in table 4.5.

Some of the latest changes to the cellular technologies allow for full rate and half rate voice service. Full rate services consume more bandwidth than half rate voice. This should result in a lower usage cost for half rate systems. Dispatch services are now starting. Dispatch services regularly allow for only half duplex operation where users can communicate only in one direction at a time. The primary advantage of dispatch service is the ability for multiple users to be simultaneously connected to each other (group call). The billing rates for dispatch services usually involve a reduced per minute rate for each subscriber that is connected to a group call.

Messaging

Currently many carriers do not charge for messaging capabilities other than the airtime charges incurred to retrieve them. These ancillary services can include both numeric and text messaging capabilities. This will no doubt change as more and more users get acclimated to receiving messages via PCS/cellular devices and carriers address solve coverage and technical challenges which currently prevent widespread use. Commonly messaging is limited to approximately 160 characters. Messages can be cascaded to create correspondence that is longer than 160 characters.

Data

Because cellular and PCS systems regularly use narrowband radio channels, they can only provide low speed data services. These data services may include circuit switched data or packet switched data. Circuit switched data is presently the most popular.

When using circuit switched data, the user typically pays only for the airtime used. Circuit switched data transfer rates on analog and digital systems are usually limited to about 9.6 kbps. For continuous data transmission of over 30 seconds, this results in a cost of less than 1 cent per kilobyte of data transferred. For short burst of data transmission, circuit switched data can cost over $1 per kilobyte of data because the call setup time is much longer than the data transmission and most cellular systems charge a minimum of 1 minute fee per call.

Because the modem protocols differ from radio, a cellular system may have a pool of modems (called a modem pool) that allows a customer to connect a modem that is optimized for cellular transmission. There is ordinarily no charge for this special connection as the cellular carrier continues to make money from the amount of airtime used.

In 1996, cellular packet services started to be offered on cellular systems. The data transfer rate for cellular packet data systems is normally limited to about 9.6 kbps. The service charges also include an account setup fee, monthly minimum charge and a usage fee that is based on the number of packets or the amount of kilobytes of user information that is transferred. The typical usage charge for packet data services ranged from approximately 1 cent to 20 cents per kilobyte.

Future Enhancements

Software Defined Subscriber Equipment

Without a doubt there exists a vision that will facilitate the development of software programmable, multiband, multimode subscriber units. Software programmable radio devices will allow a subscriber unit to adapt its modulation, multiple-access method, and other characteristics to be able to communicate with a wide range of different systems and more efficiently support diverse voice, data, image and video requirements. Like a "learnable" universal remote control, your decision for wireless service will be based on a true personal value, not technology!

Spatial Division Multiple Access (SDMA)

Spatial Division Multiple Access (SDMA) is a technology which increases the quality and capacity of wireless communications systems. Using advanced algorithms and adaptive digital signal processing, Base Stations equipped with multiple antennas can more actively reject interference and use spectral resources more efficiently. This would allow for larger cells with less radiated energy, greater sensitivity for portable cellular phones, and greater network capacity.

Figure 4.6 shows a typical 120 degree sectored antenna (an antenna which covers 1/3 of the cell site area). Within the coverage area denoted, the antenna can communicate with only one subscriber per traffic channel in the radio coverage area at any given time. The performance of the system is constrained by the levels of interference present.

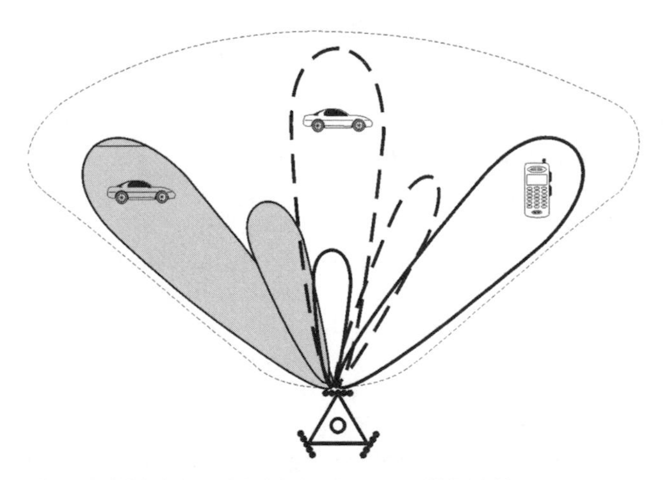

Figure 4.6 Spatial Division Multiple Access (SDMA)

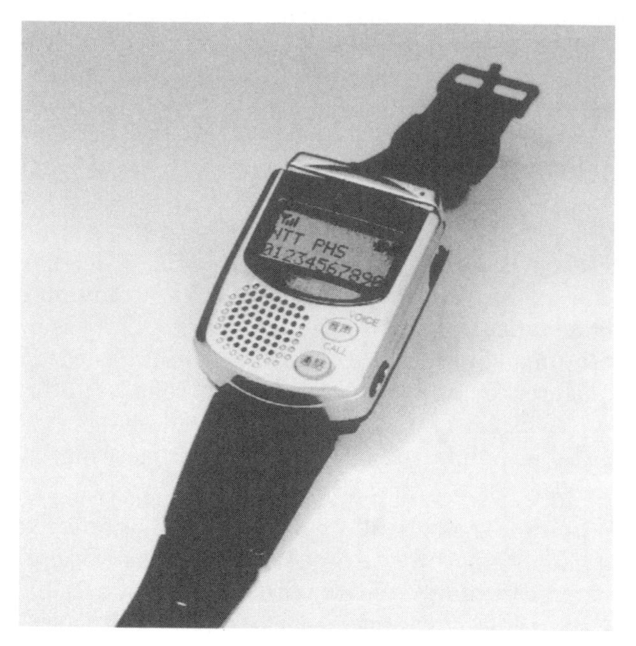

Figure 4.7 Wristwatch Telephone
Source: NTT

Wearable Communicators

Like the Wright brothers assuming a body of metal the size of a Boeing 747 aircraft could fly through the air, character artists assumed one day we would be communicating with a Dick Tracy-like wrist watch. Companies like AT&T and Nippon Telegraph and Telephone (NTT) Corporation have prototyped such wrist phones already. No bigger than a match box, fully functional, with voice activated dialing, these miniature wireless telephones are being tested today by real people and are expected to be available to the consumer market in Japan in the year 2000.

There is a technical reason why these phones will debut in Japan. Japan's Personal Handy Phone (PHP) cellular standard is a very low power system. The low power system also indicates the cells are very small and close together. By default, this system designed for densely populated conditions also requires less power in the phone for transmitting radio frequency energy. With this minimal output benefactor and Lithium Ion battery technology, the wrist telephones will feature an amazing standby time of 100 hours.

The wrist telephone is receiving the most press; however, this is only the beginning - the industry has had a goal of developing wearable communication devices for some time. The technology will quickly be incorporated into other low-power miniaturized mobile communications products.

International Mobile Telecommunications - 2000 (IMT-2000)

Combining cellular mobile voice communications and access to the information superhighway creates a myriad of opportunities for an emerging new market called the Infocom Industry. Information and communications in the palm of your hand - presents a measure of capabilities heretofore never conceived. With pinpointed applications as the key, Infocom can further distance the leader from the pack. Because the lines have been blurred by deregulation between the telecom, computer, and the media industry, the meeting-of-the-minds creating solutions outside the box has given birth to a new generation of opportunities. Opportunities to leapfrog human kind out of the Information-Age and into the Virtual Reality-Age.

In real terms, Infocom is represented by the convergence of wireless Telecommunications and the Internet. New speeds will bring wideband multimedia solutions worth investing in. A worldwide effort called International Mobile Telecommunications - 2000 (IMT-2000) is leading this (potentially 100-year) convergence.

[1] William Lee, "Mobile Cellular Telecommunications Systems", pg. 2, McGraw Hill, 1989.

[2] Personal Interview with Michael H. Sommer, President of Information Tecnologies, 8-5-98

[3]. Cellular and PCS/PCN Telephones and Systems, pgs. 377-381.

[4] William Lee, "Mobile Cellular Telecommunications Systems", pg. 5, McGraw Hill, 1989.

[5] William Lee, "Mobile Cellular Telecommunications Systems", pg. 265 McGraw Hill, 1989.

[6]. Cellular Radio, Analog and Digital Systems, Mehrotra, pg. 177.

[7]. Cellular Radio Systems, Artech House, pg. 74.

[8]. Cellular Radio, Analog and Digital Systems, Artech House, pg. 177.

[9]. Cellular Radio Systems, pg. 77.
[10]. Cellular Radio, Analog and Digital Systems, Mehrotra, Artech House, pg. 193
[11]. Cellular Radio, Analog and Digital Systems, Mehrotra, ibid, pg. 186
[12]. Cellular Radio, Analog and Digital Systems, ibid, pg. 188.
[13]. Cellular Radio, Analog and Digital Systems, ibid, pg. 193.

Chapter 5

Land Mobile Radio

Introduction

Land mobile radio (LMR) is different from Cellular and Personal Communications Services (PCS) in that it has its roots in "push-to-talk" radio systems. Push-to-talk can also be referred to as half-duplex (or simplex), whereas the half means only half of a conversation can happen at one time. This is sometimes experienced when using a household speakerphone. Full duplex, on the other hand, allows two simultaneous conversations, similar to a classic argument between a coach and a sports official. Push-to-talk communications range from a simple pair of hand-held citizen band (CB) walkie-talkies to a new breed of enhanced, full-duplex cellular-like consumer products. Licensed by the Federal Communication Communications (FCC) as specialized mobile radio (SMR), this radio service is primarily used by the public-safety sector, the industrial & construction sector and the dispatch sector like taxi-dispatch services. Though classified by the FCC as a commercial mobile radio service (CMRS), SMR is not currently classified like cellular and PCS as a common carrier telephone system.

Connection to the public switched telephone network (PSTN) by a two-way radio is usually done by incoming calls routed to a radio interface adapter. During this process, a simplex SMR radio channel is connected to a two-way regular telephone transmission line. The wired portion of the call extends the effective distance of the radio communication link. Even though a two-way wired connection is made, the communication itself remains a half-duplex.

Land Mobile Radio (LMR)

Land mobile radio (LMR) systems are traditionally private systems that allow communication between a base and several mobile radios. LMR systems can share a single frequency or use dual frequencies. Where LMR systems use a single frequency when mobile radios must wait to talk, this is called a simplex system. To simplify the mobile radio design and increase system efficiency, some LMR systems use two frequencies; one for transmit and another for receive. If the radio cannot transmit and receive at the same time, the system is called half duplex. When LMR systems use two frequencies and can transmit and receive at the same time, this is called full duplex. When a company operates a LMR system to provide service to multiple users on a subscription basis (typically to companies), it is called a public land mobile radio system (PLMR).

Figure 5.1 shows a two-way radio system. In this example, a high power base station (called a "base") is used to communicate with portable two-way radios. The two-way portable radios can communicate with the base or they can communicate directly with each other.

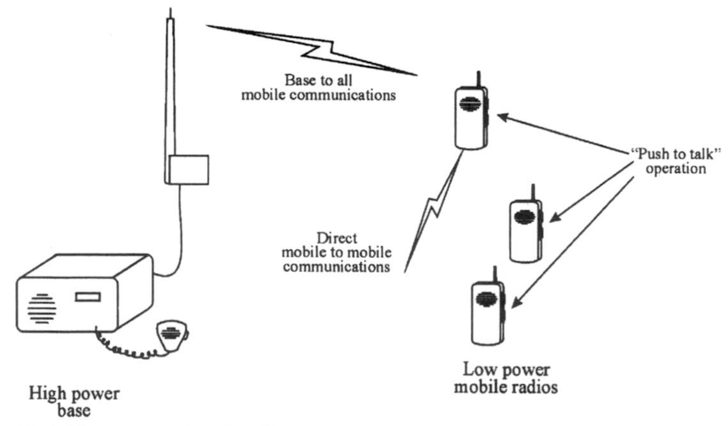

Figure 5.1 Two-way Radio System

Specialized Mobile Radio (SMR)

Specialized mobile radio (SMR) systems allow communication between mobile radios and one or more radio towers. SMR systems evolved to the allow trunking of radio channels (more efficient radio channel use). These systems use automatic radio channel assignments compared to manual frequency channel selection used in LMR systems.

SMR systems commonly provide advanced services not possible with traditional LMR systems. SMR systems always use radio channel pairs, whether or not the subscriber units operate half-duplex (requiring push-to-talk) or full duplex. Although direct mobile to mobile communications is possible (called "talk around"), it is not the usual mode of operation. Talk around is useful if the subscriber units leave the coverage area of the system, or in case of system failure. Originally, SMR systems were limited to a single tower site. Linking between systems was difficult, with roaming between systems by a given subscriber being very difficult. Telephone interconnect has always been available. More recent technological innovations have provided for advanced features such as multi-site networks, roaming capability, etc. SMR systems ordinarily cover a range of about 25 miles by using a single tower. Some SMR systems are interconnected but this is not typical.

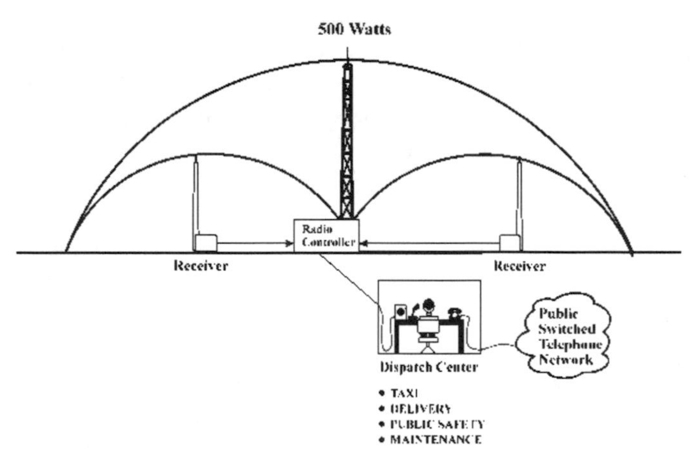

Figure 5.2 Traditional SMR System

Figure 5.2 shows a typical SMR system. In this system, one or more dispatchers communicate to mobile radios in a relatively large geographic area (normally a city). A single SMR service provider (carrier) may provide service to several taxi companies, truck delivery companies and other groups of users, so each group of users must have their own identification codes. Mobile radios communicate by coordinating with the nearest SMR tower and transmitting requests for service.

Enhanced Specialized Mobile Radio (ESMR)

Enhanced Specialized Mobile Radio (ESMR) appears almost exactly like a cellular phone system. ESMR is basically defined by who is providing the service, more so than the actual technology being employed. Older SMR systems were owned by individual entrepreneurs serving limited areas. Interoperability between systems, say, between Motorola, Ericsson/GE and EF Johnson (now Transcrypt) systems was not possible. ESMR is regularly associated with a class of SMRs that use digital radio technology to provide advanced telecommunications services to customers. These services include telephone, dispatch, numeric and alphanumeric paging, fax and data services. A good example of the new breed of ESMR is the Nextel system. From a customer's point of view, Nextel ESMR service is similar to traditional cellular service.

A key benefit of the ESMR systems is the combined push-to-talk feature along with the full duplex telephone operation packaged into one mobile telephone. The user of this combined handset has the choice of using the push-to-talk mobile-to-mobile feature or using it as a "cellular" phone. Priorities can be given to public safety (e.g., State Police, Fire & Rescue) to prevent consumer ESMR users from loading up the system during public emergencies.

Like cellular, ESMR systems have a complex of cells operating at near the original cellular frequencies around 900 MHz. ESMR calls are all routed to a Mobile Switching Center (MSC) just like that in a cellular system.

Most enhancements for the ESMR system are accomplished by digital radio transmission. There are several digital radio technologies employed for ESMR systems including TDMA and frequency hopping spread spectrum (FHSS). The leading digital ESMR systems include integrated Dispatch Enhanced Network (iDEN) developed by

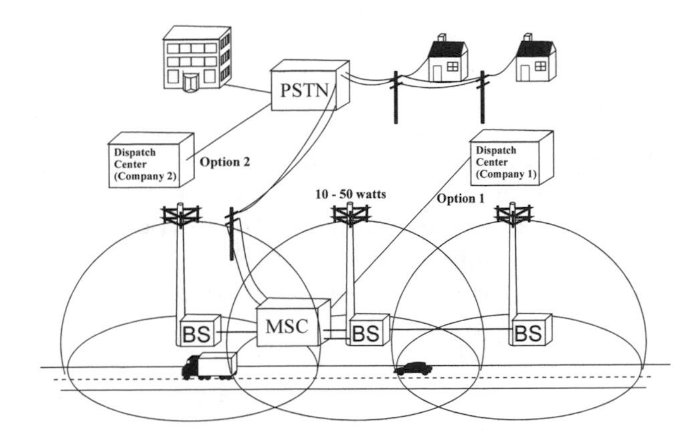

Figure 5.3 Typical ESMR System

Motorola and Enhanced Digital Access Communications System (EDACS) developed by Ericsson. These digital systems were designed to service large numbers of radio users and provide telephone, dispatch and data services that compete with traditional cellular and PCS services.

In most ESMR systems, there are many mobile radios and mobile telephones. ESMR systems have many cell sites (base stations) in a geographic region that can simultaneously transmit and receive on several radio channels at the same time. Each cell site has a radio coverage radius of about 5 to 15 miles. The systems usually provide service to large geographic regions of several cities or even nationwide service. Many are interconnected.

Figure 5.3 shows a sample ESMR system. In this system, ESMR mobile radios communicate with the system by first requesting access to the network on a control channel. The system responds to a valid request with a radio channel assignment. The ESMR system also has a switching system that can connect ESMR mobiles to each other or to other networks such as the public switched telephone network. The mobile switching center in an ESMR system must operate very quickly to allow two-way push-to-talk operation. When an ESMR customer presses the "push-to-talk" button, they must be connected to other radios in their group within about ¼ second or less. This is very different than a cellular phone system where the setup of a telephone call that can take 20 to 30 seconds.

Market Growth

The first SMR systems became operational in 1974. By 1997, there were almost 4 million SMR radio customers in the United States. Of these, approximately 50% are digital customers. Because of the ability of SMR companies to offer Enhanced SMR (ESMR) cellular-like services, this number is expected to increase dramatically.

ESMR services offer some distinct advantages over other wireless services, including no roaming charges and the ability to contact more than one person simultaneously within the network. Enhanced messaging and dispatch services round out the offering. While not aimed or attracting the attention of low usage consumers, many companies with a mobile workforce are signing up. The construction industry has especially embraced this technology. Estimates indicate that by 2006 there will be approximately 12 million users. SMR and ESMR market growth is shown below in figure 5.4.

	1997	1998	1999	2000	2001	2002
SMR	3	4.2	5.1	5.9	6.2	5.7
ESMR	1	2.1	3.2	5	6.1	8.9
TOTAL	4	6.3	8.3	10.9	12.3	14.6

Figure 5.4 SMR and ESMR Market Growth
Sources: Natwest Securities, Strategis, and APDG Research

Technologies

The technologies that are used in land mobile radio systems include basic two-way radio, trunked radio, analog and digital systems.

Two-way Radio

Two-way mobile radio systems allow communication on a dedicated radio channel between two or more mobile radios. Traditionally, two-way radio systems have served public safety and industrial applications. These systems, as a rule, only allow push-to-talk service (one-way at a time) to allow the sharing of a single radio channel frequency.

Until the early 1990's, many two-way radio system communication protocols were unique to the manufacturer of the radio equipment. Because the protocols (radio language) used in two-way radio systems have become more standardized, the selection of the type of two-way radio system is now more determined by cost, reliability, coverage distance and type of services desired.

Two-way radio systems are available in a wide range of frequency bands. Natural properties of radio waves indicate, as a rule, that lower frequencies provide longer ranges. Antenna selection and usage also effect the optimization of these two-way systems.

CB Radios operate in the 27 MHz frequency band designated by the FCC for citizens band radio. A moderately priced consumer pair of handheld walkie talkies will typically give simplex (half-duplex) voice communications over about one mile. Commercial units with high power base stations can reach ranges over 20 miles. In 1998, an estimated five million CB Radios are actively in use in the United States, causing unwanted interference on many channels. CB Radio channel 9 is reserved for emergencies (though not enforced) and most state highway patrol units monitor channel 9.

A very popular, commercial two-way service called dispatch, allows a base station operator to broadcast over the entire coverage area but only to a select group of radios. For example, ABC Taxi service uses a dispatch service, but allows only ABC Taxi cabs to hear the voice announcement. In some cases all group members hear the responding mobile, and in other cases a private response is made back to the base. These basic systems are mostly analog half duplex technology. The name two-way to describe this technology is misleading, because conversations travel only one-way at a time. The single frequency in question however, sees traffic going two ways, transmit and receive.

Because two-way radio systems allow many users to share a large geographic region, it is typical that interference can occur between users that are operating close to each other. If a user would receive all conversations on the radio channel, it would be annoying. For this reason, squelch systems were developed. Squelch is a process where the audio output of a radio receiver is controlled (enabled) by the reception of an incoming RF signal that is above a predetermined level. Squelch allows a radio user to avoid listening to noise or interference signals of distant radio transmissions that occur on the same frequency. Squelch systems can be carrier, tone or digital code enabled.

Carrier controlled squelch systems mute the audio until the incoming radio signal is above a specific signal level. Carrier controlled squelch allows a radio user to avoid listening to noise or weak interference signals of distant radio transmissions that occur on the same frequency. Unfortunately, any strong incoming radio signal will pass through the squelch system. This includes nearby users whose conversations may not be of interest to the user. To overcome this limitation, tone controlled squelch systems were developed.

Tone controlled squelch systems mute the audio of a radio receiver unless the incoming radio signal contains a specific tone. Tone controlled squelch allows a radio user to avoid listening to noise or interference signals of all other radio transmissions that occur on the same frequency that do not have the correct tone mixed in with the audio signal.

Digital squelch systems are very similar to tone controlled squelch systems. Digital squelch systems mute the audio of a radio receiver unless the incoming radio signal contains a specific digital code. Digital controlled squelch allows a radio user to avoid listening to noise or interference signals of all other radio transmissions that occur on the same frequency that do not have the correct code mixed in with the audio signal.

Figure 5.5 shows a typical two-way SMR radio system. In this diagram, a single radio channel is shared by several mobile units that operate within the radio coverage limits of a high power radio base

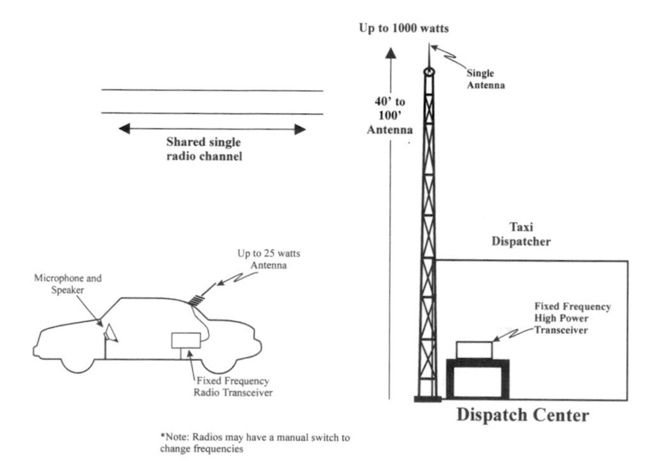

Figure 5.5 Two-way SMR Technology

station. To help differentiate between different groups of users, a separate squelch tone signal is mixed in with audio signals for specific groups of users (for example, a taxi company and a flower delivery service). This squelch tone is ordinarily a frequency that is below the audio frequencies so the users cannot hear the tone.

Trunked Radio

Trunked radio is a system that allows mobile radios to access more than one of the available radio channels in that system. If a radio channel is busy, the mobile radio can access another radio channel in the system that is not busy. This type of system allows mobile radios to overcome blocked channels during busy periods.

Many of the newer SMR technology groups have converted their dedicated radio channels into a group of radio channels (called a trunk group). Mobile radios that operate on trunked radio systems can choose any radio channel that is unused at the moment. This technique greatly decreases the amount of blocked calls compared to the use of dedicated radio channels. Most trunked SMR equipment still operates only in the half-duplex mode, which requires a push-to-talk operation by the SMR user.

Figure 5.6 shows a typical trunked SMR system. In this example, there are several available radio channels. Mobile radios that operate in this system that wish to communicate may search for an available radio channel by looking for identification tones. Optionally, some

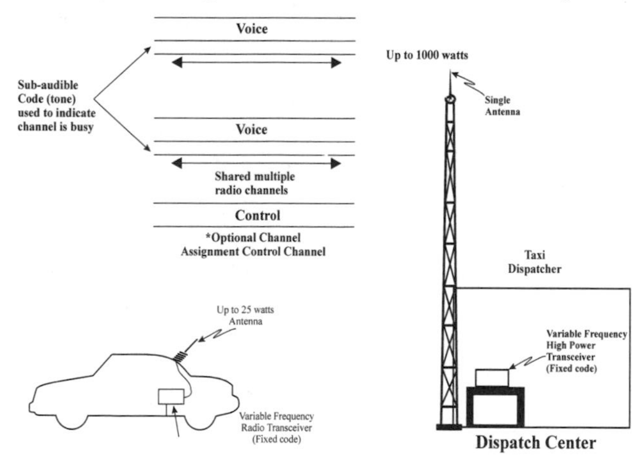

Figure 5.6 Typical Trunked SMR Technology

trunked radio systems use dedicated radio channels to coordinate access to radio channels.

Analog

Many of the traditional two-way radio systems use analog radio technology. Most of these systems use FM modulation, however there are some systems that use single sideband amplitude modulation. Single sideband modulation reduces the amount of frequency bandwidth required for each radio communications channel. Until recently, single sideband AM modulation suffered from radio distortions. New modulation techniques have overcome many of these limitations and equipment is now available that uses single sideband technology.

Figure 5.7 shows a typical analog two-way radio system. In this diagram, a mobile radio has an FM transmitter and FM receiver bundled together. Because the mobile radio is used for "push-to-talk" service, the antenna does not need to be connected to both the transmitter and receiver at the same time. When the user presses the push to talk button, the antenna is connected to the transmitter. When the button is released, the antenna is connected to the receiver. A squelch circuit is connected to the receiver to allow the receiver audio to reach the speaker only when the receiver level or tone code is of sufficient level to ensure a good received signal. When the level is below that threshold (normally set by the user via a squelch knob), the speaker is dis-

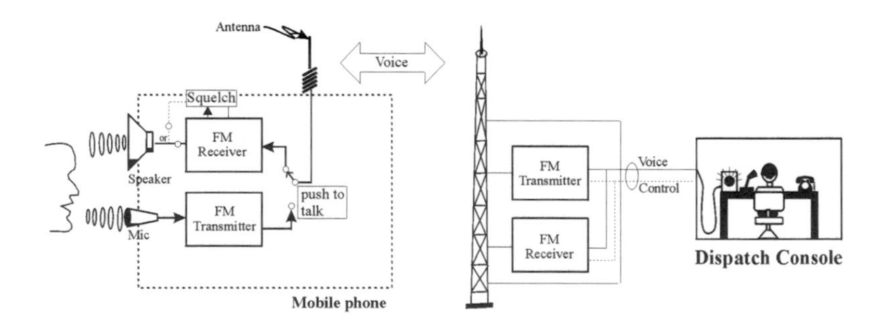

Figure 5.7 Analog Two-Way Radio

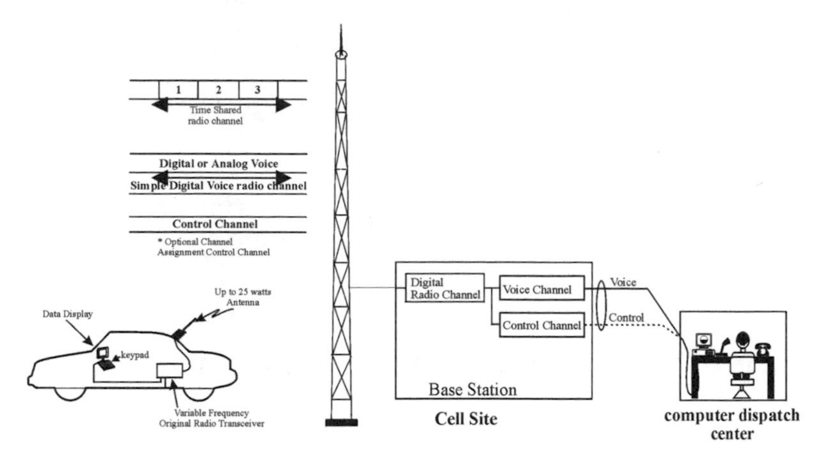

Figure 5.8 Digital Land Mobile Radio

connected and no sound (such as noise) can be heard. The mobile radio communicates with a radio tower that has a high power base station. The base station usually is connected to a control console that allows a dispatcher to communicate or to patch the audio channel to another location (such as a telephone line connection).

Digital

Many of the newer two-way land mobile radio systems use digital radio technology. Most of these systems use some form of phase modulation. The use of digital transmission allows for many new advanced services.

Figure 5.8 shows a typical digital two-way radio system. In this example, a digital mobile is connected to a data display in a mobile vehicle. The radio transceiver has a transmitter and receiver bundled together. Some digital systems combine the control channel with a traffic channel while others use dedicated control channels. Digital systems can either use a single digital radio channel for each user or can divide a single radio channel into time slots for use. The use of time division allows for several users to simultaneously share each radio channel.

Most digital land mobile radio systems can provide for short message services (SMS) and data transfer. These services allow numeric and alphanumeric paging and data transfer. Enhanced digital systems allow for handoff from one radio channel to another. This allows for cellular like operation. To allow for advanced services (such as messaging and radio channel handoff), control messages and data may be sent between the base station and the digital mobile. Control messages may command the digital mobile to adjust it's power level, change frequencies, or request a special service (such as three way calling). To send control messages while the digital mobile is transferring digital voice, the voice information is either replaced by a short burst (blank and burst) message, in some systems (also called fast signaling), or else or else control messages can be sent along with the digitized voice signal (called slow signaling).

Commercial Systems and Standards

There are over 100 types of analog and digital land mobile radio systems in use. Some of these systems use protocols that are unique (custom) to the manufacturer and customer. Some of the more popular systems or standards in use are integrated dispatch enhanced network (iDEN)®, enhanced digital access communications system (EDACS), MPT1327, TETRA, Passport, and APCO 25.

Integrated Digital Enhanced Network (iDEN)®

Integrated digital enhanced network (iDEN) is a digital TDMA ESMR system. The iDEN system was developed by Motorola. iDEN is available in the United States and over 13 countries throughout the world. There are over 16 network operators that use iDEN technology that include Nextel Communications Inc., Southern Communications (Southern LINK) and ClearNET Communications (Canada). Most of the United States and Canadian population is covered by iDEN service.

iDEN service is used for integrated cellular like service, paging, data and dispatch services. Key attributes include combined voice and messaging, large geographic coverage region and dispatch services.

Formerly, iDEN was called the Motorola integrated radio system (MIRS). It was developed for the company called FleetCall, prior to it becoming Nextel. The system first offered commercial service in 1994 in California. Since its introduction, the iDEN technology has evolved to provide higher quality voice service (cellular like) and cost effective dispatch voice and messaging.

The iDEN system only uses one type of digital radio channel. This radio channel is 25 kHz wide and has a gross data rate of 64 kbps. The radio channel is divided into frames that have 6 time slots. This allows up to six users to simultaneously share a single radio channel. To increase the voice quality for cellular like operation, two slots are combined per frame to allow three cellular voice users per radio channel. The modulation type is quadrature amplitude modulation (QAM). This allows the 25 kHz channel to provide an efficient 2.56 bits of data for each Hertz of bandwidth.

The RF power of mobile radios is 6/10ths of a Watt during the transmitter burst. The recent addition of advanced sleep modes to the iDEN technology provided for much longer standby time battery life.

Enhanced Digital Access Communication System (EDACS)

Enhanced digital access communications system (EDACS) is a combined analog and digital SMR system that provides for voice, dispatch and data trunking radio service. EDACS was developed by Ericsson. It is used in the Americas, Asia and Eastern Europe. These systems are primarily used for (public safety, industrial, etc) fast data and messaging access.

The present digital channel only allows one user per channel. It is anticipated that the evolution of the EDACS system will allow multiple users to share each radio channel via TDMA channel division. EDACS radio channels use 12.5 kHz or 25 kHz radio channels to provide 4800 or 9600 bps data. EDACS uses a dedicated control channel to coordinate system operation. The control channel operates at 9600 bps.

The common frequency bands for the EDACS system include 150, 450, and 800 to 900 MHz ranges. The EDACS digital channels provide for voice privacy using encryption schemes. The data channel signaling on EDACS allows for late entry of group members.

The EDACS allows for automatic power level control and mobile radios can be produced with output power up to 5 Watts. While talk around is possible with EDACS, it is not commonly used. The key attributes of EDACS systems include a fast radio channel access time and the ability to deploy multi-site networks (e.g. state wide systems).

Ministry of Posts and Telegraph 1327 (MPT1327)

MPT1327 is an analog trunked private land mobile radio standard that is primarily used in Europe, Asia and developing countries in other parts of the world. The MPT1327 standard can be used to implement systems with only a few radio channels (even single-channel systems) to large interconnected networks. The MPT1327 system was developed to standardize land mobile radio equipment and services. The system is primarily used for public safety applications but it can be used for cellular like services.

The system has two types of radio channels; control channels and the traffic (voice or data) channels. Control channels can be dedicated or non-dedicated. Dedicated control channels are permanently available for sending and receiving control information. Non-dedicated control channels can be dynamically converted to a traffic channel. This as a rule occurs if all the other channels are in use.

Radio channels are 25 kHz wide. They have a data signaling rate of 1200 bps and the modulation type is Fast Frequency Shift Keying (FFSK) subcarrier modulation. It is designed for use by two-frequency half-duplex radio units and a duplex base station. The system can have up to 1024 channel numbers and 32768 system identity codes.

The MPT1327 system allows for voice, dispatch (group call), data, emergency and messaging services. Messaging services allow up to 184 bits of text data to be sent between units or to the control center. A unique MPT1327 service is the sending of up to 32 status messages that can be sent between units. Some of these messages are predefined while others can be specified by a user.

Passport

PassPort is an enhanced trunking protocol designed for wide area networking. PassPort was developed by Trident Micro Systems, a subsidiary of Trident Datacom Technologies, Inc. PassPort systems are in use throughout the world. PassPort is primarily targeted to business, industrial and commercial SMR applications. The PassPort protocol and NTS system development started in 1994 with the first commercial deployments of the technology in 1997.

The PassPort protocol is downwardly compatible with analog LTR systems. It supports auto registration and de-registration as radios move between sites. It operates on the NTS networked trunking infrastructure platform developed by Trident Micro Systems. The NTS is a unique distributed networking environment that does not require a central or hub switch with all sites in the network connecting to a common point. The NTS platform provides fast network call set up times. A scaleable switch is located at each RF site. Sites may be connected in any combination of a star, linear, mesh or hub network configuration.

The key attributes for PassPort include wide area dispatch networking, voice mail for interconnect and dispatch, and ESN security. Additionally, NTS is an all digital platform and provides an easy migration path for operators between analog RF and digital RF when such product becomes widely available.

All radio channels in a PassPort system are available for voice. All channel assignments are conducted sub-audibly leaving the in-band portion of the signal free for voice. No channels are set-aside for control purposes.

The radio channel bandwidths it is currently in use in are the 25 kHz and 12.5 kHz applications. Data rates (control, signaling) in PassPort signaling are identical to LTR. The modulation type PassPort is a trunking protocol designed primarily for FM. The NTS platform however, is designed to support high speed future digital signaling formats and digital RF.

APCO Project 25

Associated public safety communication officials (APCO) project 25 is an digital trunked land mobile radio standard that is primarily used in the United States. The standard was accepted in 1993 and APCO Project 25 compliant systems are primarily used for public safety applications.

One of the key objectives for the Project 25 standard was backward compatibility with standard analog FM radios. The Project 25 standard allows for a simple migration from analog to digital Project 25 systems. This permits users to gradually replace analog radios and infrastructure equipment with digital Project 25 systems.

There are two radio channel bandwidths used in the Project 25 system; 12.5 kHz and 6.25 kHz. The 12.5 kHz radio channel uses (C4FM) modulation and the 6.25 kHz radio channel uses (CQPSK) modulation to achieve a more efficient data transmission rate. Because of the modulation types selected, the typical receiver is capable of demodulating either the C4FM and the CQPSK signals. The digitized voice uses a 4400 bps improved multiband excitation (IMBE) voice coder. This is the same voice coder type selected by INMARSAT, for use in satellite communication devices.

At the start of every transmission, there is a header word. These header words have heavy error correction. Headers are followed by voice frames that contain signaling (control information) and encryption protected user data for voice privacy.

Each header word is preceded by a synchronization word (to alert that a new header is coming), a network identifier code. The header word includes an encipherment (encryption) type and code(s), and an destination address code (if necessary). The Project 25 systems allows multiple encipherment key codes.

Voice frames follow the header word. Voice frames are 180 ms in length and pairs of voice frames compose a 360 ms Superframe. Part of the voice frames are dedicated for link control. If the addressee is a talk-group, the voice frames contain information that describes the type of information type of message to follow (group call, data, etc..), manufacturer identifier, priority indicator, talk-group address, and the transmitting radio's identifier. All members of a talk group can receive the transmissions of all other members of that talk-group. Because each voice frame has a group address and type information, this allows members of a talk-group to enter the group call after it is established (late entry). The digital radio channel uses forward error correction and interleaving to allow operation in when the bit error rate is up to 7 percent.

A low-speed data channel is provided in the digitized voice frame structure (88.9 bit/s). This data channel could allow applications such as accurate geographic location information or measurement of channel quality information.

Terrestrial Trunked Mobile Radio (TETRA)

TETRA is a digital land mobile radio system that was formerly call Trans European Trunked Radio. The TETRA is being developed by the European Telecommunications Standards Institute (ETSI) to create a more efficient and flexible communication services from both private and public-access mobile radio users.

TETRA is capable of sending and receiving short data messages simultaneously with an ongoing speech call. It effectively supports voice groups and has capacity for over 16 million identities per network (over 16 thousand networks per country). TETRA permits direct mode operation (talk around) that permits direct communication between mobile radios without the network. TETRA includes a priority feature to help guarantee access to the network by emergency users. The system allows independent allocation of uplinks and downlink to increases system efficiency. The signaling protocol supports sleep modes that increases the battery life in mobile radios.

The TETRA system is fully digital and allows for mixed voice and data communication. It is specified in open standards. The TETRA system allows up to four users to share each 25 kHz channel. It allows interworking with other communication networks via standard interfaces. TETRA is capable of call handoff between cells and it has integrated security (user/network authentication, air-interface encryption, end-to-end encryption).

TETRA mobile radios have the option to use subscriber identity module (SIM) card for security keys and personal data. TETRA systems have an Inter-System Interface (ISI) that allows interconnection of TETRA networks from different manufacturers and gateways provide access between the TETRA network and other networks. Line Station interface permits the connection of third party dispatch systems.

TETRA data rate (kbps) include unprotected, standard and high protection. Unprotected data rates are 7.2, 14.4, 21.6, 28.8, data rates for standard protection (kbps) 4.8, 9.6, 14.4, 19.2. and data rates for high protection 2.4, 4.8, 7.2, 9.6. The system is capable of sending predefined status messages (range of over 32,000 values). User messages variable in length up to 2047 bits.

Tetra operates in the VHF and UHF frequency ranges of 150MHz to 900MHz. The RF carrier spacing is 25 kHz which is divided into 4 communication channels multiplexed onto each carrier (TDMA). The modulation method is phase shift keying (Pi/4 DQPSK). The channel data transmission rate: 36kbps which provides a net data throughput: 28.8kbps (7.2 kbps per channel). The digital voice coding is ACELP.

The base station transmitter power levels include 0.6W, 1W, 1.6W, 2.5W, 4W, 6.3W, 10W, 15W, 25W, 40W and the mobile radio is capable of 1W, 3W, 10W, 30W. TETRA base stations operate in full frequency duplex. Mobile stations may operate in frequency duplex or half duplex. The TETRA standard supports continuous, timeshared, and quasi-synchronous operation.

Services

Voice

Mobile radio service through SMR and ESMR systems operates similarly to cellular systems. The most well known application for land mobile radio is wireless voice communications. Voice communication can be basic telephony or two-way voice (radio to radio). The service rates for land mobile radio voice applications are typically lower than cellular systems. For private systems where the radio license is owned by the customer, there are no service fees aside from maintenance.

Activation Fee	$25
Monthly Access Charges	$45 (includes 250 minutes of usage)
NOTE: Both Caller and Receiver	on Network are charged
Additional Airtime Charges	$0.10 each additional minute
Telephone Connect Charges	$0.05 for each minute

Figure 5.9 Typical ESMR Service Rate Plan

When SMR operators provide service for public use (such as for a taxi cab company), a flat monthly fee is charged and there is no usage fee. When the land mobile radio system offers telephony service, a fee is commonly charged for connection to the public telephone network, since operator must pay for the wired telephone line (and sometimes the usage of the line). ESMR systems ordinarily charge per minute usage fees for both two-way voice and basic telephone service. Common ESMR service rates are shown in figure 5.9.

Dispatch

Dispatch radio service normally involves the coordination of a fleet of users via a dispatcher. All mobile units and the dispatcher can usually hear all the conversations between users in a dispatch group. Dispatch operation involves push-to-talk operation. Common dispatch service rates are shown below.

Activation Fee	$0.00
Monthly Access Charge Per Mobile Unit	$15 (includes unlimited amount of usage)

Figure 5.10 Typical Dispatch Service Rate Plan

The commonality between dispatch service and mobile telephony is that communication occurs for brief periods. There is usually no charge for the airtime usage for SMR systems, but ESMR systems commonly charge usage fees for dispatch service. Figure 5.10 shows a typical rate plan for dispatch type service.

Paging and Messaging

Paging and messaging services allow users to send and receive text messages up to approximately 180 characters per message. Messages can be cascaded to allow the sending of longer messages.

Depending on the type of system, the charge for unassisted messaging services varies from "no cost" when bundled with other services (such as voice) up to 50 cents per message. The per message cost ordinarily comes from operator assisted messaging.

Data

One of the fastest growing areas for land mobile radio is data service. Data services transfer information between computers and data terminals. Examples of data devices used in land mobile radio systems include computers used in police cars and digital dispatch terminals used in taxi cabs and delivery trucks.

A new type of operator console and service has been created to take advantage of data services on LMR systems. This service is called computer aided dispatch (CAD). CAD systems are a computerized communication system that can coordinate and/or track mobile vehicles. CAD systems can comprise various degrees of complexity, from automated messaging devices to complex computer systems that display maps and vehicle positions on a computer monitor.

Similar to the messaging fee structure for LMR service, the fees for data services varies from "no charge" to 10 cents per kilobyte.

Future Enhancements

Packet Data

There are several high speed packet data services in development (100 kbps+). These high speed data services will allow land mobile radio companies to serve more customers and support new types of applications, such as sending pictures, that are unsuitable for low speed data messaging.

Dual Technology Mobile Radios

There are some developments in progress which combine other wireless technologies with land mobile radios. These include mixing GSM service with SMR radio technology or satellite service with SMR radios. This will allow advanced services and permit international roaming.

Chapter 6

Mobile Wireless Data

Introduction

Mobile wireless data is the transmission of digital information, through a wireless network, to radios that typically move throughout a geographic area. The success of mobile wireless data systems had been limited, until the mid 1990's, due to the high cost of service and equipment, and limited geographic coverage areas. In the mid 1990's, service usage and equipment costs dropped due to competition and availability of applications and end user devices. Many wireless data systems are now available globally.

In addition to the mobile wireless data systems covered in this chapter, there are also many types of fixed wireless data systems. These systems include wireless local loop (WLL), wireless local area network (WLAN) and wireless cable. The fixed systems are covered in chapter 9.

The growth of the Internet has also enabled low cost, standardized access to wireless data networks which is accelerating the growth of the wireless data marketplace. In 1998, some wireless data systems had more then 700,000 active devices operating in their system. It is

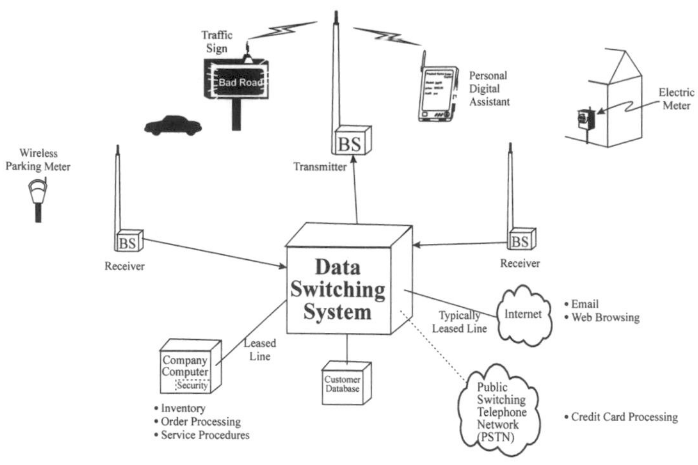

Figure 6.1 Wireless Data System

predicted that there will be over 10 million wireless data devices in use by the year 2001.

Figure 6.1 shows a basic wireless data system. In this example, many types of wireless data devices communicate through a public wireless data system. In the core of the system, there is a switching system. The switching system commonly routes the data between the wireless device and a computer system (such as a company computer). In this diagram, there are more receivers than transmitters. This is required to allow low power mobile data transmitters to reach the system. Base station transmitters can provide up to 500 Watts effective radiated power (ERP) while portable mobile data devices can usually provide less than 1 Watt of transmitted power.

Market Growth

Overall Market Growth

With the demand for high data rate communications solutions, paralleling interest in the Internet, (fueled by easy-to-use application software, its wide array of text, graphics, video and audio content), data market growth has increased substantially. The availability of Internet services over wireless radio channels will be a critical factor in determining overall market growth.

To date, most wireless data applications are non-human in nature. These include applications such as monitoring wireless parking meters, vending machines and environmental concerns among others. Human access includes the ability to access data available on the Internet, private intranets, new services and e-mail. The Internet, for example, is being used by businesses for building interactive branding via communication with customers, advertising products and services, publishing product specifications; and acting as a source for point of sale applications.

Market growth for wide area wireless data communications services is in the early stages, primarily because wireless data is not yet capable of providing high data transfer rates at a cost comparable to fiber optic cable or wired connectivity. However, the overall market growth of the wireless data market is up. In 1997, there was over 21% growth for circuit switched data (primarily cellular data) and over 89% growth for packet data (ARDIS, RAM, CDPD, and Ricochet).

The question that is raised by this is "Do people REALLY require high speed wireless access? The quick answer is "yes" because people have been programmed to always want the fastest connection possible. But, considering the cost model of wireless vs. wireline, is it possible that relatively slow access would suffice if cost and coverage were superior? Most people don't really want (that is, don't want to pay for) high-speed wireless web browsing. Instead of a high-speed connection at a premium price, a slower speed connection can be provided a much lower cost. Figure 6.2 reflects the market growth rate for wireless data applications.

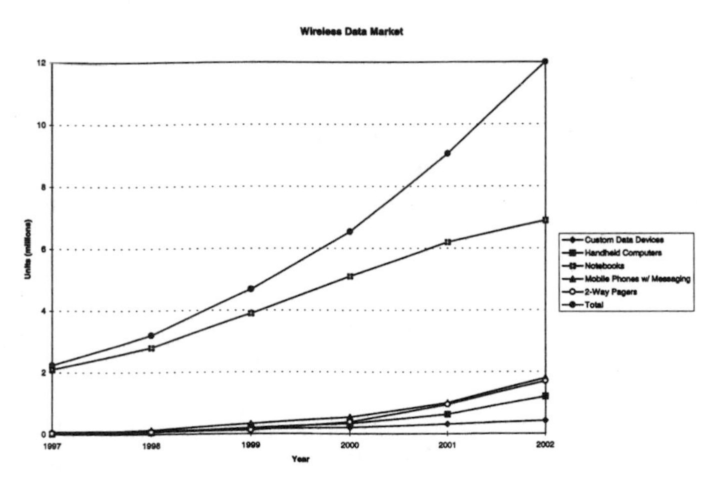

Figure 6.2 Wireless Data Market Growth

Technologies

There are several wireless technologies that are used for mobile wireless data. These technologies allow data rates to vary from a few kilobytes per second to several hundred kilobytes per second.

Wireless data technology is commonly divided into two types of transmission: circuit switched data and packet switched data. Circuit switched data is characterized by the continuous connection between two points. Packet data is identified by the dividing of data into small packets that may take different routes between points.

There is a new type of wireless data called "control channel data. Control channel technology utilizes the excess capacity of a wireless system control channel to transmit packets or burst of data. The technology is usually limited to small amounts of data.

Circuit Switched Data

Circuit switched data is the continuous transfer of data between two points. To establish a circuit switched data connection, the connection address (normally a phone number) is sent first and a connection path is established. After this path is setup, data is continually transferred using this path until the path is disconnected by request from the sender or receiver of data.

Circuit switched data would be understood better if it was entitled "A phone-call for data" because a connection (usually a public telephone call) must be made beforehand. After the phone call is connected through the public *switched* telephone network (like most modems do on home computers), the telephone *circuit* remains in conversation mode while data is transmitted back and forth between the two devices. For circuit switched data, the telephone circuit resources (wires and switching) remain dedicated to (and typically paid for by) the user regardless if data transmission has stopped.

Figure 6.3 shows a circuit switched data system. In this figure, a computer is sending a data file through a radio channel to a home office computer. To start the data file transfer, the computer uses its internal radio modem to dial the phone number (dialed digits) of the office

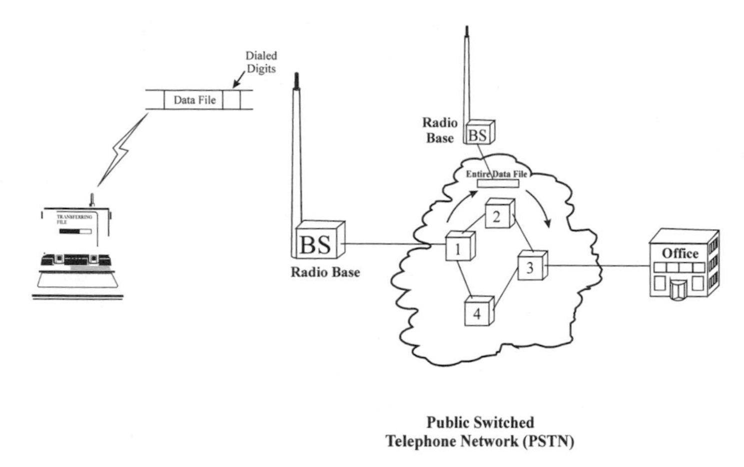

Figure 6.3 Circuit Switched Data

computer. The dialed digits are received by the radio base station which routes the call to the home office through the public switched telephone network. In this example, the call is routed through switches 1, 2 and 3 in the network. As soon as the connection is made, the computer can start sending data to the office. Throughout the connection, this path will be maintained without any changes.

Packet Switched Data

Packet switch data is the transfer of information between two points through the division of the data into small packets. The packets are routed (switched) through the network and reassembled at the other end to recreate the original data. Each data packet contains the address of its destination. This allows each packet to take a different route through the network to reach its destination.

Packet data is ideally suited for information services that have short bursty (i.e., the data transfer occurs only during a small amount of the time – in "bursts" – while most of the time there is no data transfer required) communication requirements. The primary advantages of packet data transmission include rapid connection time and efficient use of resources for short data transmission requirements. The disadvantages of packet systems include inefficiency due to the inclusion of an address message with each packet of data and unpredictable delays associated with transmission.

Unlike circuit switched communication which keeps a communication channel in use regardless if data information is temporarily halted (such as browsing the web), packet data systems only transmit when there is information to be sent. The packet data system does not maintain a constant connection between the two users. This type of system is referred to as "connectionless" because there are no pre-determined time periods or dedicated resources for packet transmission.

Packet switched systems divide a customer's data information into small packets that contain their destination address. Packets are sent to their destination by the best path possible at the time of transfer. The travel time for each packet between its origin and destination may be different. This is because packets of information are often sent on different routes due to communications path availability. As packets are received, they are reassembled in the proper order at the receiving end.

Figure 6.4 shows a packet switched data network. In this example, a credit card machine will send data through a radio channel. First, the digital information is captured from the credit card. The credit card machine then sends the digital information to a packet radio modem. The modem divides the digital information into small packets (a single packet can commonly hold about 100 characters) and adds the address to the beginning of the data message. The data message is then sent through the radio channel to a radio data base station. The radio data base station sends these packets into the packet switching network which routes packets through available paths towards its

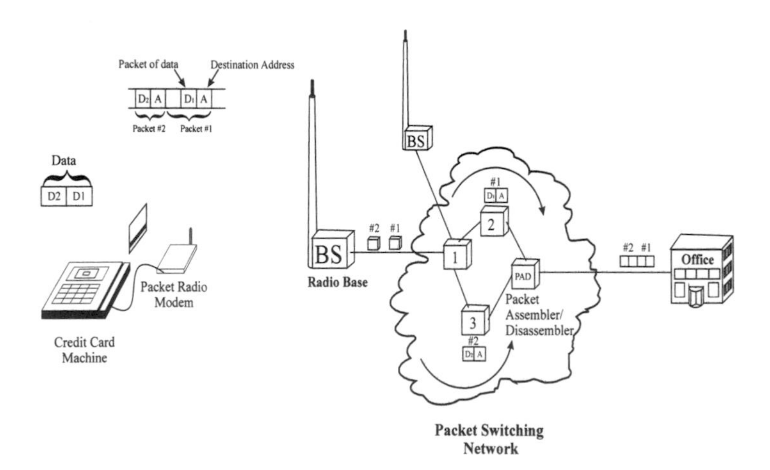

Figure 6.4 Packet Switched Data

destination. When it reaches its destination, the packets are recombined in a packet assembler/dis-assembler (PAD) and supplied to the computer in the office which processes the credit card data.

The question of choice between circuit switched data and packet switched data is often an easy one, but may not be predicated totally upon technological capabilities. In the applications where communication is continuous and loaded with data transfer - circuit switched data is the choice. An example of an appropriate circuit-switched application would be continuous data file transfer. Connection to email is an example of an application that is well suited to packet switched data, because the majority of time is spent standing by to send or receive email. Field service dispatch and courier pick-up & delivery are also examples of when packet data would be the appropriate application choice. All use short bursty messages, one or two packets out and in per transaction. In the real world, even though an application may be suited for a certain technology, availability, standards compliance or good marketing campaigns often make the decision for the public.

Research has shown that wireless data applications are solutions oriented, not technology based. The solution may be a combination of price, technology, and application, and the bottom-line decision will not be based on technology. The solution should solve a specific problem.

Recent reductions in pricing for wireless data service combined with easier interfacing with the internet have drastically altered the price based decision making process between circuit and packet. Some carriers have announced "all you can eat" wireless data service plans which make the decision to go with packet much easier to justify. With circuit data being charged by the minute and packet data offering a fixed rate, the primary decision becomes not price but coverage.

Sub-Band Data Transmission

Sub-band wireless data transmission involves the sending of data signals with other radio signals, such as radio or television broadcast channels, through a radio control channel . Sub-band data transmission takes advantage of the unused capacity of a radio channel and the large geographic coverage area of high power radio stations. One of the key advantages of sub-band or control channel data transmission is the limited amount of system investment. Most sub-band data transmission systems only require an update to the switch or network

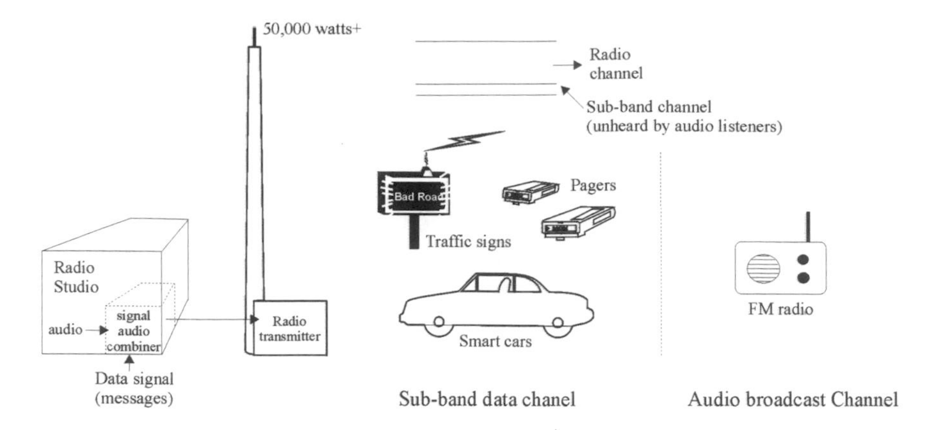

Figure 6.5 Sub Band Data System

head end. Once this is accomplished, the control channel data service is available throughout the area serviced by that system. There is no additional requirement to upgrade each radio transmission tower equipment.

Figure 6.5 shows how a radio channel is used to send data signals. .In this example, a data signal is combined with a radio broadcast studio signal and they are sent to the high power radio transmission tower. The sub-band audio signal is shifted in frequency so that listeners to the audio broadcast signal do not hear the sub-band data signal. Devices such as pagers, road signs and smart cars have special receivers that can separate the data signal from the audio broadcast signal.

Shared Voice and Data

Shared voice and data systems allow data transmission on radio channels when voice communication is not in process. The concept of shared voice and data allows a voice system to offer data services during idle radio channel periods.

Figure 6.6 below shows the basic operation of a shared wireless data system network. The wireless data device scans radio channels (step 1) for a free one. After it has found and locked onto a free radio channel, it will begin to transmit data (step 2). If the radio channel is interrupted by another activity (such as a voice signal), the wireless data device will re-tune to the next available wireless data radio channel

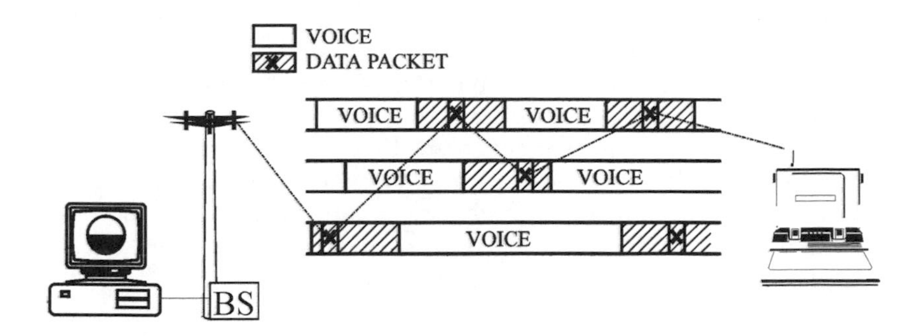

Figure 6.6 Shared Voice and Data System

and continue to send data. This technique treats the radio channels that have voice activity (voice sessions) as noise. This is an example of the frequency hopping technique described earlier. The important point here is that voice is treated as the superior application. If voice has the channel, the wireless data system looks elsewhere for an unoccupied radio channel. If the channel is quiet, the data hops on. If data is active and another voice session comes up, the data hops again. This is possible since the data is operating in packet mode. The packets can be reassembled at the receiving end no matter which air channel is used.

Wireless Modem Protocols

Modem protocols are the language that modems use to communicate with each other. Modems convert digital signals (normally from a computer) into audio signals that can be sent through a network (usually the telephone network). In addition to the conversion of user data signals to audio signals, modems communicate with each other using protocol language.

The protocol language used by landline modems is ordinarily different than the protocols used by wireless modems. Landline modem communication protocols are optimized for communication paths that usually have consistent quality or a limited number of bit errors. Radio modem protocols are optimized to adapt to the different types of distortions that may occur on a radio channel. Typical radio modem protocols include ETC/ETC2, MNP 10, TX-Cell and EC^2.

It is possible for standard modems to communicate through radio channels (such as cellular). However, if the radio channel experiences distortion, this will result in difficulty transmitting and may even result in a lost data connection. When using a radio channel for data communication, it is best to use modems with the protocols that have been optimized for radio transmission. Even with the use of modems that are specifically designed for radio transmission, a realistic wireless data transfer rate (throughput) is approximately 9600 bps. Cellular capable modems regularly have 28,800 or even 56K stamped all over them, but those rates are only apply when the modem is being used on a wireline connection [1].

Modem Pools

Modem pools are the inter-working functions between a mobile radio modem and a wired telephone modem. Modem pools contain equipment that convert radio modem protocols to wired telephone modem protocols. A wireless system must coordinate the selection of which modems will be selected in the modem pool.

A wireless system may include modem pools that adapt wireless data modem protocols to standard telephone protocols. To access a modem pool, a mobile radio must first connect to the wireless network and request a modem for digital service. For some systems, such as the GSM system, this is a message that is sent from the mobile phone that requests for the call to be processed as a data call through a modem. For other systems that cannot send such messages (such as analog cellular networks), the user must dial a special phone number that routes the call to a modem pool. After the connection is established, the mobile phone then dials the destination number for the modem to be connected. After the connection is established, the radio modem communicates with the modem in the modem pool. The modem in the modem pool then converts the information to the selected landline modem format.

Distributed Data Collection and Distribution

To overcome the challenge of interconnecting large numbers of radio base stations, some systems use distributed data collection and distribution to interconnect base stations without wires. In this type of system, the mobile data radios (end points) communicate with nearby base station hubs which gather their information. These hubs redistribute to other base station collection points which ultimately reach a data collection and distribution center. This allows the rapid deployment of a wireless data network without the added cost of interconnection cabling.

In a distributed system, the end radios (e.g., utility meters) periodically transmit to a small base station that is mounted on a nearby telephone pole. This base station temporarily stores the information and forwards it on to another nearby base station that is mounted on a building. This base station is connected to a computer system that gathers and processes the data. By using this distributed data collection system, base stations can be rapidly deployed without requiring wired interconnections.

Internet Connectivity

The Internet uses Transaction Capability Protocol/Internet Protocol (TCP/IP), which is the fundamental language of the Internet. While TCP/IP is a universal standard for Internet communications, it uses a large portion of the data transmitted for signaling. Because the amount of available data in a wireless data system is usually limited and has higher costs associated with it, many wireless data systems do not directly use TCP/IP. They usually have protocol adapters that allow the conversion of their proprietary protocol to TCP/IP protocol. This allows most wireless data systems to offer interconnection to the Internet (while keeping the airlink usage to a minimum).

Wireless Data Applications

Wireless data is used for hundreds of applications. Most wireless data applications are primarily narrow vertical market applications that offer specific solutions. Wireless data services vary from low data rate mobile applications to high speed data for fixed location applications.

Applications can be divided into **vertical** (industry specific) and **horizontal** (mass market) types. Historically, wireless data services have been used for the business community in vertical applications. In the mid 1990's, the cost of wireless data service and equipment drastically dropped, nationwide service became available and the Internet growth helped some wireless data applications to become horizontal.

Application	Technology
Wireless point-of-sale credit card devices	*Packet*
Parking Meters	*Packet*
Railroad Crossings, traffic statistics	*Packet*
Telemetry, water levels, temperatures	*Packet*
Vehicular Location Tracking and Data Collection	*Packet*
Utilities	*Packet*
Dispatch delivery messaging	*Packet*
Wireless E-mail	*Packet*
Wireless Internet	*Circuit*
Mobile Office Applications (Microsoft CE...)	*Circuit*
Wireless Fax	*Circuit*
Field Service	*Circuit or Packet*
Sales Force Automation	*Circuit or Packet*
Hand-held Computing	*Circuit or Packet*

Figure 6.7 Wireless Data Types and Applications

Horizontal applications include wireless email, Internet browsing and interactive paging applications.

There are many applications for wireless data. These include security and fire alarm service monitoring, personal safety devices, remote status monitoring for vending, postage machine monitoring, bill to coin change machines, commercial washing machines, office and factory equipment, and intelligent energy control devices such as remote control thermostats, intelligent transportation monitoring systems for traffic lights, parking meters and toll booths and hundreds of other applications. Figure 6.7 shows some typical applications for wireless data and the types of data transmission that are as a rule used.

Credit Card Devices

Credit card devices are commonly referred to as wireless point of sale credit card verification. These devices operate away from power and land-line connections but have the capability to receive credit card information (swipe card) and transmit the it for validation and processing.

While the ability to use wireless credit card machines allows mobile vendors to accept charge cards without high risk, wireless credit card systems can also be used in fixed locations. This eliminates the requirement of using local telephone lines and in some states, eliminates the usage costs of those telephone lines (called message units).

Parking Meters

In 1997, there were approximately 4.5 million parking meters in use in the United States, generating over $4 billion in annual revenue. Some of the revenue comes from coins deposited and other revenue comes from ticketing for expired meters. It is estimated that less than 30% of the potential revenue is collected. Over $10 billion is lost because of unmonitored, expired meters. Wireless parking meter monitors alleviate many of the high costs associated with meter monitoring. Parking meters that have wireless transmitters can call an attendant when it detects it has run out of time and a car is in front of it (by a sonic sensor).

Estimates show that enforcement personnel have been able to write approximately three to six more tickets per hour with the wireless parking meter system in place. An officer usually spends about seven hours per day issuing parking tickets. Therefore, assuming an average increase of about four more tickets per hour, he/she can issue roughly 28 tickets more per day. The national average is about $15 per ticket, and ranging from as low as $10 per ticket to as high as $40 per ticket. At $15 per ticket, that amounts to approximately $420 per day per officer. According to the National Parking Institute, on average, each enforcement personnel monitors 200 meters. This translates into a potential additional revenue of roughly $2.10 per day per meter, or a monthly additional revenue of $63.00 per wireless parking meter [2].

Vending

Remote monitoring of vending machines reduces the number of delivery visits, determines the optimal amount of inventory to be carried on the delivery trucks and reduces the cost of servicing the vending machines.

Vending machines are typically located in remote areas without any communications connection. Even in urban locations, high installation and service costs prohibit connecting vending machines to a dedicated wire line. Wireless remote vending combines the benefits of point-of-sale credit card authorization, inventory control, and advertising. Vending machines that can authorize and accept charge cards or other types of money cards make it possible to sell more merchandise without requiring the customer to have enough or the exact change. Wireless remote vending also informs distributors of inventory levels in the machines, removing guesswork from ordering and dispatching. Finally, vending machines can be a good outlet for advertising.

Environmental Monitoring

Wireless service is the most practical method of transferring small amounts of telemetry and monitoring information in rural areas. Environmental monitoring applications include water levels, earthquake sensors, fire watch (temperature), contamination level monitoring, bridge corrosion detection and other types of sensors that are located in rural regions.

Energy Management

Energy management can be enabled by wireless networks. Smart "thermostat-like" panels which allow businesses and homes to use electricity more efficiently by programming heating, ventilation and air conditioning (HVAC) system and hot water heater. Devices within the building would communicate with one another over existing electrical wiring using power line carrier (PLC) technology. Two-way wireless data networks could provide a connection between the building and a utility to deliver pricing signals, home management services, and public information, and to send customer messages and device status signals back to the utility.

Dispatch

Dispatch is the monitoring and scheduling of information to workers. It consists of issuing work tickets, providing routing information or directions, tracking and data collection, inventory levels and order entry for parts and other mobile work related functions. Wireless data provides many companies, such as taxi and limousine companies, with dispatch messaging services.

Utility Monitoring

There are approximately 230 million electric, gas and water meters in the United States [3] and many are potential candidates for network meter reading (NMR). In 1998, there were are over 770,000 electric meters that were monitored by a wireless network in the United States [4].

Utility industries are undergoing a fundamental and broad-based transition to a competitive marketplace. Utility system operators must regularly track power flow and billing data. Wireless data transmission systems can do this in a more timely and cost effective manner by eliminating the need for manpower, overhead, and operation costs, especially in remote areas.

Wireless E-mail

Wireless email is the sending and receiving of short messages of text to a portable computing device. Because most wireless data systems have limited data transfer rates and have a high cost of service for large amounts of data, a wireless data system must allow the user to select if the delivery of large attachments is desired.

Wireless Internet

Wireless Internet involves communicating with the Internet, particularly the web, without wires. Because communicating with the Internet can require the transfer of large amounts of data (for pictures and audio clips) and interconnection has been normally costly, wireless Internet has had limited success until recently.

There are several advantages associated with wireless Internet. The first is the Internet service provider can commonly bypass the local telephone company. Users of their service may directly connect to their computers without the need for a telephone line. Next, wireless Internet service provides more efficient use of the Internet site communications interface. This is because a single interface can be processing multiple users as opposed to the Internet service provider having modems for each user that is connected to a phone line.

To allow more efficient communication between wireless devices and the web, handheld device markup language (HDML) was created. HDML provides similar features as hypertext markup language (HTML) without the intense use of graphics, through the limited use of input keys. This allows devices such as cellular telephones to take advantage of information on the web without transferring large quantities of data.

Mobile Computing

Mobile computing provides the ability for mobile workers to connect to information sources while on the road. Mobile computing can involve a combination of wired and wireless connection. Typical applications for mobile computing include sales force automation, inventory tracking and scheduling appointments and tasks.

Advertising

Advertising is starting to find its way into wireless devices. Already, some cellular phone systems automatically call a roaming customer that has entered into their system to inform them that they are authorized to use the service in their system.

Wireless data can be used to send advertising messages to billboards, displays on vending machines or to set top boxes that display text or graphics on televisions or monitors.

Vehicle Monitoring Services

In some vehicles, wireless devices such as cellular telephones and vehicle tracking systems are being integrated as part of the vehicle's electronics system. Vehicular applications include traffic alert bulletins and re-routing directions. Some automobile manufacturers have begun the standardization effort to create plug-and-play equipment for the consumer electronics in automobiles. In one application, if the car develops a fault (breaks down), a wireless data transmitter automatically calls for help to the nearest location and begins transmitting stored computer data to aid in diagnosis of the problem which is likely occurring.

Commercial Wireless Data Systems

Data over Analog Cellular

To transmit data over analog systems, a computing device (ordinarily a portable computer) is connected to a wireless modem. The modem converts the digital signals from the computer to the audio tones for transmission through the radio channel. Sending data through an analog cellular network is independent of which type of analog cellular system is used.

Data Over Digital Cellular and PCS

For digital cellular systems, a modem is not required, however a data adapter is required. While it is possible to directly transfer data files via a digital telephone, there are few phones that have this direct data connection capability at this time. By the year 2000, this feature will be widely available in all the current airlink technologies, CDMA TDMA and GSM.. With a circuit switched data connection, the cellular or PCS system must convert a digital data signal to an analog signal so it can be sent through the public switched telephone network (PSTN). This digital-to-analog conversion in the cellular switch is done by modems that convert the digital wireless signals into analog modem signals that are transferred through the PSTN. Digital data is available today only on GSM networks. CDMA and TDMA systems will offer this capability soon.

AMERICAN MOBILE'S ARDIS NETWORK

The ARDIS network is a public packet data system developed by Motorola. ARDIS originated from the private network dedicated to IBM's service personnel in the 1980's. ARDIS commercial service started in January 1990 in the United States. The original ARDIS technology used the MDC4800 protocol which had a data transfer rate of 4,800 bps. This has evolved to use the RD-LAP protocol which allows a data transfer rate of 19,200 bps. ARDIS service covers approximately 430 metropolitan areas in the US. As of March 31, 1998, the ARDIS network is owned and operated by American Mobile Satellite Corporation.

The original ARDIS technology used the MDC4800 protocol which had a data transfer rate of 4,800 bps. This has evolved to use the RD-LAP protocol which allows a data transfer rate of 19,200 bps. ARDIS service covers approximately 430 metropolitan areas.

In 1998, ARDIS had 1700 base stations and planned to add radio transmitters for increased capacity, and radio coverage. (twenty-five service areas have been added) According to ARDIS, their existing system capacity can handle approximately 1 million customers.

The network of base stations are interconnected by a supported ARDIS backbone network. This wireline portion of the network provides the communications infrastructure between a customer service, the ARDIS Service Engine, the Radio Network Controllers and the Radio Base Stations.

The ARDIS Network uses single-frequency re-use (multiple base stations to the same receiver) to ensure good in-building coverage. Unfortunately, this simulcast system does not increase the system capacity. ARDIS' wireless data network uses licensed radio channels (Specialized Mobile Radio Service) that operate in the 800 MHz frequency band.

Mobitex/RAM

Mobitex is a public packet data system developed by Ericsson. Mobitex is available in countries throughout the world. In the United States, Mobitex technology is deployed by RAM. Commercial service started in October 1990. Recently, Mobitex system began to offer two way paging service.

RAM uses a frequency reuse system which allows for frequency handoff in a similar process that cellular does. Because Mobitex systems use frequency re-use, Mobitex systems have the ability to increase their capacity through the addition of more base stations.

Mobitex systems ordinarily operate in the 400 MHz and 800 MHz bands. Mobitex systems typically operate in the 400 MHz and 800 MHz bands. A Mobitex radio channel transmits at a rate of 8 kbps and has a bandwidth of 12.5 kHz. This allows the transmission of 2,800 to 13,000 packets per hour. Each packet contains up to 512 bytes and status messages. As an example, an application which requires ten medium packets (66 bytes) per hour allows a maximum of 800 users per channel. Mobile data radios usually transmit up to 2 Watts peak and the output power from a base station is 6W or lower and is adjustable in 3dB steps.

A Mobitex network is organized as a hierarchy with nodes at three levels: base stations (BRS), area exchanges (MOX) and main exchanges (MHX). At the top level there is also a Network Control Center (NCC). Radio coverage is provided by overlapping radio cells, each served by an intelligent base station. Because intelligence is distributed throughout the network, data packets need only be forwarded to the lowest network node common to the sender and receiver. The base station handles all local traffic between mobile terminals so only billing information is passed up to higher levels.

Several Mobitex radio modems allow the use of data protocols that include an extended AT command set. This makes them compatible with existing communications software that runs on standard PCs.

Cellular Digital Packet Data (CDPD)

Cellular Digital Packet Data (CDPD) is a wireless data transmission technology developed for use on cellular phone frequencies (824-894 Mhz). CDPD systems started service in the United States in 1994 and are deployed in several other countries in the Americas and China.

CDPD uses cellular radio channels to transmit data in packets when the channels are not being used for voice communication. The CDPD system is a shared voice and data system. This technology offers data transfer rates of up to 19.2 Kbps, better error correction and quicker call set up than using modems on an analog cellular channel.

CDPD has achieved radio coverage of over 65% of major metropolitan areas in the United States, primarily because of its "piggy back" on the analog cellular networks. Many consumer based products offer CDPD end user accessories for connectivity to the wireless world. Because CDPD is deployed with cellular, which has 10 times the number of base stations that ARDIS and RAM have, CDPD generally has much better radio signal coverage and has very good mobility features such as invisible handoff between cells. The data connection will be maintained throughout the coverage area. If the MES travels in a blocked area such as a tunnel, the data connection is re-established once back in active coverage.

CDPD does not always have coverage comparable to cellular voice because CDPD requires an update not only to the switch, but to each cell site as well. In some areas, the cellular carrier has elected not to offer CDPD throughout the voice coverage area due to cost issues [5].

The user's "packet-data" ready laptop or other device is called a Mobile End Station (MES). An antenna site, "Base Station", is called a Mobile Data Base Station (MDBS). The data switch or router is called a Mobile Data Intermediate System (MD-IS).

Cellemetry[sm]

Cellemetry radio is a control channel messaging system that was developed by BellSouth to allow small data packets of information to be carried through a cellular system's control channel. The first Cellemetry systems were operational in 1996 in the United States.

The Cellemetry system sends data through the control channel by replacing the electronic serial number (ESN) that is normally sent by cellular phones during system access. The Cellemetry system replaces the ESN with user data. It uses the control channel in a way that only software upgrades are required by the cellular carrier to implement the data transfer. Because the Cellemetry system uses a standard cellular radio technology, analog mobile telephones can be modified at low cost to provide Cellemetry messaging service.

The Cellemetry system works by the periodic registration of a Cellemetry phone with a cellular system. When a standard cellular phone registers with the cellular system, it sends both its mobile identification number (MIN) along with an ESN. The MIN is the phone number that identifies where the phone is registered and the ESN is normally used to validate the identification of the MIN. The ESN holds 32 bits of information. The Cellemetry system replaces the ESN with user data such as a meter reading value. When the Cellemetry phone registers with the system, the MIN is used to locate the Cellemetry home system via the standard intersystem signaling network (IS-41) that is used to link all the cellular systems together throughout the United States. Both the MIN and ESN are sent to the home system for validation. The Cellemetry system knows from its subscriber database that the MIN is from a Cellemetry phone so it stores the ESN information as data.

Aeris Microburst

MicroBurst is a sub-channel messaging and small data packet communication system developed by Aeris Communication. It operates over the control channels of an analog cellular network. The first Aeris Microburst system started operation in 1998.

The Microburst system uses the control channel in a similar way to the Cellemetry system performs. Microburst technology requires that no cellular system upgrades are required by the cellular carrier. MicroBurst data packets, of up to approximately 50 bits, are transmitted using the standard control channel protocol and are routed automatically to a central, nationwide network host.

CellNet

The CellNet system is a wireless packet data system that operates on both licensed and unlicensed radio frequency bands. The company CellNet developed its technology and primarily provides telemetry network meter reading ("NMR") services to electric, gas and water utility companies. The CellNet network normally uses small radio-device meters, that are part of the utility meters, to monitor and periodically report usage data from each meter to a nearby radio base. As of December 31, 1997, CellNet had over 770,000 meters that were generating monitoring revenues [6].

The CellNet system has a two-tiered wireless network that permits distributed data collection and distribution points to allow rapid deployment of the network with minimal interconnection costs. The network hierarchy is managed by a central system control center which collects, concentrates, forwards and manages data from many fixed endpoints through various network levels that are all interconnected by radio. The parts of this communications system include the following: Endpoint devices which transmit data relating to the equipment they are monitoring or controlling. MicroCell Controllers ("MCCs") which manage the endpoint devices in their local coverage area (as part of a local area network or "LAN") and which collect and process data transmissions from such endpoint devices. CellMasters, which gather data from MCCs located in a wide coverage area (as part of a wide area network or "WAN"), and which communicate that data to a central System Controller. And a System Controller which manages the entire network and operates the application gateways for integration with the client's own data systems.

Endpoint devices communicate with MCCs using the 902-928 MHz unlicensed radio frequency band. The CellMaster generally coordinates 50 to 200 MCCs over an area usually covering approximately 20-75 square miles (2.5-5 mile radius). The MCCs communicate with the CellMasters using the 928/952 MHz band. CellNet radio channels have 25 kHz or 12.5 kHz channel bandwidths with full duplex operation and point-to-multipoint data services. A single CellNet radio channel can be divided up into 19 subchannels from a single 25 kHz channel.

Teletrac

The Teletrac systems provide automatic vehicle monitoring (AVM) by using transmitters that operate in the 902-928 MHz frequency band. Teletrac provides vehicle location and messaging services, primarily for vehicle theft management. The first Teletrac system started operation in the United States in 1990.

The Teletrac system consists of several transmitting and receiving antennas installed around an urban area to provide reference signals for a Teletrac receiver to calculate its location. Teletrac uses time of arrival measurements on received signals from four or more antennas to compute a vehicle's location with an accuracy of about 100 feet. Teletrac transceivers are priced around $400, and the monthly subscription charge averages about $20 per vehicle.

Ricochet

Ricochet is a wireless packet data system that uses the unlicensed radio spectrum frequency band. Ricochet networks use a wireless data communications infrastructure to provide wide area coverage in metropolitan areas. Ricochet provides wireless Internet service for a low, flat monthly subscription fee that permits unlimited usage. Ricochet service began in September 1995, and in 1998, was available in the San Francisco Bay Area, the Seattle and Washington, D.C. metropolitan areas, parts of Los Angeles, and in a number of airports, corporations, and university campuses.

The Ricochet system provides end users with data rates from 10 Kbps to 40 Kbps, depending on factors such as geography and network usage. The primary elements of a Ricochet network are that they are compact and inexpensive network radios that are deployed on street lights, utility poles and building roofs in a geographical mesh pattern.

The Ricochet radio system uses frequency hopping spread spectrum digital packet-switched radio technology which transmits in the unlicensed frequency spectrum of 902 MHz to 928 MHz. Ricochet plans to provide service in the 2.3 and 2.4 GHz frequency bands which should increase end user speeds to 128 Kbps [7].

Ricochet networks employ intelligent packet-switched technology. After a data packet is transmitted by a Ricochet wireless modem, it is routed through one or more network radios wirelessly to a wireless access protocol (WAP) where it is routed to its destination over the wired backbone. Ricochet network radios are intelligent and they communicate with neighboring network radios to automatically learn their identity, location, how well they can communicate with each other and the frequencies where they can be found at any particular point in time. When the learning process is complete, each network radio automatically routes data packets through the network. Ricochet modems can also directly communicate with each other without accessing network radios, provided that they are close enough to establish a direct radio connection.

FM Sub-band Signaling

FM High Speed Sub-Band Signaling systems send data through FM radio channels in combination with other audio broadcast information. There are several radio stations that offer FM sub-band communication. The data rates of FM range from 1200 bps to over 10 kbps. For example, CUE Network Corporation (a US corporation) offers FM sub-band communication over a nationwide satellite radio network using the subcarrier capacities of nearly 600 radio stations.

Services

There are three basic services offered by wireless data systems: circuit switched data, packet switched data and messaging.

Circuit Switched Data

Circuit switched data is a bearer service as it only transports the user's data between points. When sending data through a circuit switched connection, the user regularly pays a standard per minute charge for the amount of time that the connection is maintained regardless of how much data is sent through the channel.

It usually takes approximately 10 to 20 seconds to establish a circuit switched connection on a wireless network. This is due to the processing of dialed digits through the telephone network and the amount of time the modem requires to establish which communication language will be used (called training time). The user ordinarily pays for this setup time even if they only have a very small amount of information to send (such as an email message).

Once a connection is established on a circuit switched connection, data transfer rates generally range from 9600 bps up to 28,800 bps. Figure 6.8 shows that this results in an average cost of data transmission that is below 0.5 cents per kilobyte.

When a wireless carrier offers modem pools to the wireless data customers (approximately 3% to 4% of cellular customers occasionally transmit data), there is usually no charge for modem pool access.

Activation Fee	35
Per Minute Airtime	35 cents
Average Data Rate	9600 bps (1.2kilobyte/sec) or 72 kb/min
Per Kilobyte Cost	0.5 cents per kilobyte

Figure 6.8 Typical Wireless Data Cost over Cellular

Packet Data

Packet switched data is also a bearer type of service as it only transports the users data between points. When sending packets of data through the network, the user normally pays only for the amount of data or number of packets that they send.

Unlike circuit switched data, the connection time for packets is ordinarily under 1 second (some systems may be below 150 msec) and the user does not pay for this setup time. Figure 6.9 shows that the standard price for packet data transmission ranges from approximately 4 cents to $1 per kilobyte. A one time activation fee is as a rule required along with a minimum monthly fee. The usage amount is normally applied to the monthly fee.

Several wireless data service providers in the United States now offer service based on application and number of units. This results in different price plans that can vary from $15-25 per month per unit with some systems offering a flat fee for a fixed or unlimited amount data transmission. The trend is to move away from the per packet charge. Figure 6.9 shows a typical wireless packet data rate plan.

Activation Fee	$45.00
Monthly Access	$50.00 (includes 1000 kilobytes of data)
Additional Data Charges	8 cents per kilobyte

Figure 6.9 Typical Wireless Packet Data Service Rate Plan

Wireless Messaging

Wireless messaging is a teleservice as it processes the user data. Wireless messaging services include store and forward and Internet connectivity. Typically wireless messaging is combined (bundled) with wireless data service (such as packet data). Figure 6.10 shows a typical wireless email fee structure.

Activation Fee	$45.00
Monthly Access	$49.95 (includes unlimited access)
Coverage	Includes national coverage

Figure 6.10 Typical Wireless Email Service Rate Plan

Future Enhancements

Many of the future enhancements involve increasing the data transmission rates and the ability to send multiple types of media (voice and pictures) over a single network.

High Speed Packet Data

Similar to land mobile radio systems, there are several high speed packet data services in development (100 Kbps+). These high speed data services will allow wireless data service providers to serve more customers and to support new types of applications that are unsuitable for low speed data messaging, such as sending pictures.

Multi-Media

Multi-media is the mixing of voice, data and video services. Digital information is a universal medium capable of transporting digital voice, data files, and digitized video. High speed wireless data transmission systems add multi-media capabilities.

Satellite Data

Satellite data involves the sending and receiving of data in large geographic areas (hundreds or thousands of kilometers wide) to and from an individual or group of receivers. Satellite data services include both circuit switched and packet switched data services. Satellite data service is particularly well suited for broadcast data services such as news or other information services that are delivered to many people in a wide geographic area. For more information on satellite data services, refer to chapter 8.

1. Personal interview, Tyler Proctor, Zsigo Consulting, August 2, 1998.
2. "Wireless Parking Meter Networks," Keary Warner & Rob Gehring, Zsigo Wireless, 7 June 1997.
3. US Security and exchange commission 10k report, "CellNet", 31 Mar 1998.

4. Ibid.
5. Personal interview, Tyler Proctor, Zsigo Consulting, August 2, 1998.
6. US Security and exchange commission 10k report, “CellNet”, 31 Mar 1998.Ibid.
7. US Security and exchange commission 10k report, “Metrocom”, 31 Mar 1998.

Chapter 7

Paging

Introduction

Paging is a method of delivering a voice or data message, via a public communications system or radio signal, to a person whose exact whereabouts are typically unknown by the message sender. Users usually carry a small paging receiver that presents a numeric or alphanumeric message displayed on an electronic readout; alternatively, messages could be sent and received as voice messages or other data.

Paging has become popular among the consumer market. Previously reserved for "on-call" technicians, it is now used by some parents to call their children to dinner. The success of pagers is likely to continue because pagers normally: cost less than other devices; have longer battery life (several months); work well inside buildings; and can be extremely small, allowing easy portability.

There are four basic types of messaging services offered by paging systems: tone, numeric, text (alpha) and voice. These messaging services can be delivered by two types of paging systems: one way and two way paging. One way paging systems only allow the sending of messages from the system to the pager. Two way paging systems allow the confirmation and response of a message from the pager to the system as well.

One-Way Paging

One-way paging is a process where paging messages (signals) are sent from a radio tower to a pager without a return verification signal. In its simplest form, a one-way paging system can serve up to several hundred thousand numeric paging customers.

Figure 7.1 shows a one-way paging system. In this diagram, a high power transmitter broadcasts a paging message to a relatively large geographic area. All pagers that operate on this system listen to all the pages sent, paying close attention for their specific address message. Paging messages are received and processed by a paging center. The paging center receives pages from the local telephone company or it may receive messages from a satellite network. After it receives these messages, they are processed to be sent to the high power paging transmitter by an encoder. The encoder converts the pager's telephone number or identification code entered by the caller to the necessary tones or digital signal to be sent by the paging transmitter.

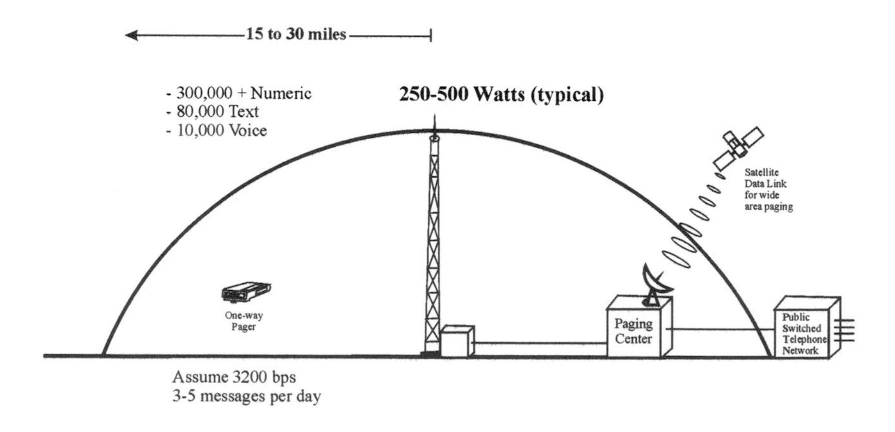

Figure 7.1 One-Way Paging System

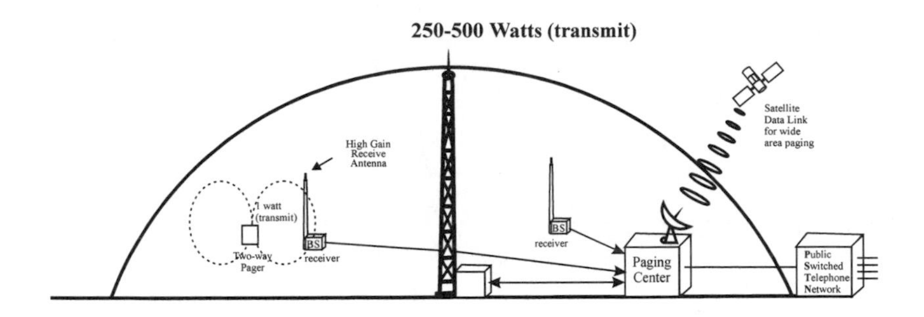

Figure 7.2 Two Way Paging System

Two-Way Paging

Two-way paging systems allow the paging device to acknowledge and sometimes respond to messages sent by a nearby paging tower. The two-way pager's low power transmitter necessitates many receiving antennas being located close together to receive the low power signal. Figure 7.2 shows a high power transmitter (250-500 Watts) which broadcasts a paging message to a relatively large geographic area, as well as showing, several receiving antennas. The reason for having multiple receiving antennas is that the transmit power level of pagers is much lower than the transmit power level of the paging radio tower. The receiving antennas are very sensitive, capable of receiving the signal from pagers transmitting only 1 watt.

The number of receivers required for a two-way paging system is dependent on the available transmit power from the pagers and how fast the information is to be transferred. The higher the data transmission rate, the higher the number of required receivers.

The main advantage of two-way paging systems is their ability to require pagers to register their location within the paging system. This allows the paging system to direct pages for a specific pager only to the area near where the pager last registered. This frees up the paging capacity of channels in other geographic areas so paging messages can be sent to other pagers. This is a type of frequency reuse based on geographically separated systems.

Different "pagers" use different technologies to operate. Even if they operate on the same frequency band, different pagers may use different communication protocols. This can result in a pager from one company not being capable of working on another company's network. It is not the visible features like Alphanumeric Text Display, or buttons, the difference is in the air-interface protocol.

Narrowband PCS

Narrowband PCS is 3 MHz of bandwidth in the 900 MHz band that is used for two-way messaging (paging) services in the United States. The Narrowband PCS frequency channels were auctioned off by the FCC in 1995. Narrowband PCS is different from Broadband PCS which allows for two-way simultaneous voice as well as data communications.

Market Growth

End-users who at one time were content to receive a message via a simple "beep" are now able to receive weather, stock quotes, news and even e-mail via pagers. This, in conjunction with paging form factors and service pricing dropping to consumer levels over the last ten years, has fueled tremendous growth within the industry. With the number of paging customers in the United States rocketing from approximately 2.6 million in 1984 to approximately 63 million in 1998, and estimates predicting total US pagers in use reaching over 101 million by the year 2001, it is clear that consumer level paging products and services have been embraced with open arms.

1997	1998	1999	2000	2001
52.7 million	63.2 million	74.6 million	86.6 million	101.3 million

Figure 7.3 U.S. Paging Units in Use
Source: WinterGreen Research, Inc.

For example, in 1996, 35% of paging customers were consumers, while 65% were commercial users. Consumer usage has shot up 21 % since 1993. By 2003, the consumer market is expected to reach approximately 51 % of the total market.

Product pricing has certainly drifted lower over the years, as mentioned, contributing to consumer growth. In 1992 the average price of a numeric pager was $126 and the average price of an alpha pager was $191. In 1997 numeric pager pricing averaged almost 54% less, at $58. Alpha pagers dropped an average of just over 30%, down to $132 in 1997.

One-way paging growth will continue until 2001, when two-way growth will begin to erode the one-way market. Two-way subscriber growth is estimated at 500,000 in 1997 to 23 million in 2002 by some estimates. Two-way data subscriber growth is expected to go from an estimate of 150,000 in 1997 to over 10 million in 2002.

Alpha paging is also showing strong growth. Alpha paging subscribers numbered less than 2 million in 1994 and are expected to increase to over 10 million by the year 2000. Wireless e-mail subscriber growth is also expected to increase from less than 1 million in 1996 to over 16 million in 2001.

Sources: PCIA, WinterGreen Research, Strategis Group, MMTA, and APDG Research

Technologies

Some of the peripheral technologies that have developed have also facilitated paging services. These technologies include interactive voice response (IVR), localized paging, queuing, pagers with synthesizers, simulcast service, paging protocols and increases in system capacity through a combination of these technologies.

Interactive Voice Response (IVR)

Early paging systems required the use of operators to key in the identification code of pagers to be notified of an incoming page. Over the past few years, paging systems have evolved to use interactive voice response systems that guide a caller through the paging process. This reduces the operational cost (reduces staffing levels) and normally increases the reliability of the paging system to capture and send paging messages.

Localized Paging

Localized paging is the sending of messages only to transmitters that are located near where the pager may be located. The need for localized paging comes from the desire to use pagers throughout a large geographic area. Because one-way systems do not transmit back to the paging system, there is no way to know where the pager is or if the message even arrived at the pager. This requires the same message to be transmitted to paging transmitters in all areas where the pager may be operating. This means that one-way paging systems that offers national service must transmit the page over the entire nation. This means that a particular frequency, or frequency channel is simultaneously sent to its nationwide network of antennas of 500 to 1,000 transmitters.

To increase the capacity of one-way paging systems that cover large geographic regions, some systems with multi-city or national coverage require the user of a pocket pager to call in via telephone when they travel, and identify their location. This allows the radio paging to be limited to that one city where the pocket pager is located and thus avoid impact on the paging traffic elsewhere in the country.

Two-way paging systems know the location of pagers. This allows localized paging where the paging system can send paging messages only to the systems or paging towers where it knows pagers are operating. If a pager begins to operate in a new system, it will register with the paging system. The paging system will then inform the home paging center of the system location of the pager. The paging center stores the pager location in a location register. The next time the paging system receives a page for that pager, it can send the page only to that system.

Queuing

One of the main advantages of paging compared to mobile voice service is its non-real time messaging. Because paging messages can be delayed up to several minutes, this allows messages to be placed on a waiting list (queued) when the paging system becomes busy.

Pagers with Synthesizers

One of the key changes was the production of pagers that use frequency synthesizers. In the past, pagers used crystals that allowed them to operate on a single frequency. If the customer wanted to change paging companies or if the paging system became busy on a specific frequency, pagers would need to be "re-crystalized." Pagers that use frequency synthesizers can be programmed by software to change frequencies. This eliminates the need for crystals. The only significant disadvantage is synthesizers cost more money than crystals.

Simulcast

To grow the system coverage and increase quality, paging operators added antennas and used a simulcast radio transmitting technique. Paging antennas are located every 10 to 50 miles, creating a huge coverage area. A simulcast transmission system uses multiple transmitters operating on the same frequencies to provide radio coverage throughout a larger geographic area. The primary benefits of simulcast transmission include better in-building penetration, better coverage, and lower cost pocket paging receivers. The disadvantage of simulcast transmission is that the same signal is sent on each of the radio towers. This does not increase the system serving capacity.

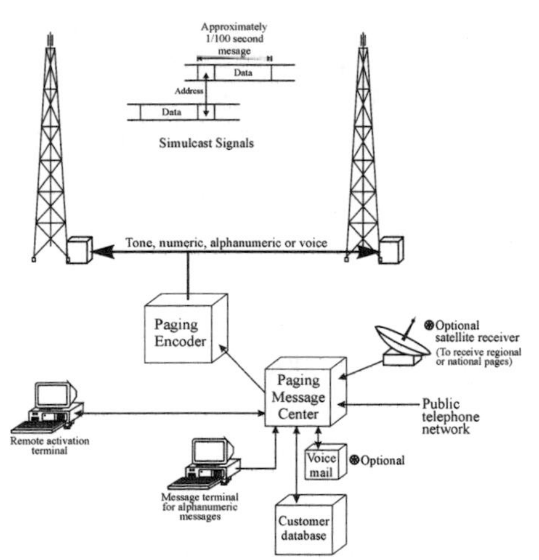

Figure 7.4 Simulcast Paging

Figure 7.4 shows a paging system that uses simulcast transmission. In this example, the same paging message is sent to two paging transmitters. As can be seen in this example, the challenge with simulcast paging is as the pager is closer to one tower then the other, the transmit delay time can cause the signals to not directly overlap. Because radio signals travel so quickly, this delay is minor. However, it can result in some dead spots due to signal adding or subtracting.

Paging Protocols

Paging protocols are the language used to communicate between the radio tower and the pager. Protocols define how pagers are addressed and how the messages are sent on the radio channel. Protocols differ in their data rates, maximum number of pages that can be addressed in a system, and types of services (tone, voice, text or numeric). The typical data rates transmitted on paging systems vary from approximately 1200 bps to 6400 bps. Some new paging systems can transfer data up to 112 kbps.

There are several different protocols used for paging systems. Some paging protocols can be mixed on the same radio channel. For example, POCSAG protocols can be sent on the FLEXTM system.

System Capacity

The capacity of paging systems is primarily determined by the data transfer rate of the radio channel and the amount of data that is required for each message. For tone only pagers, only the address of

Data Rate (bps)	*Numeric*	*Alpha*	*Tone + FM Voice*	*Digtal Voice*
512	50000	12000	1500	1500
1200	120000	30000	1500	3750
1600	160000	40000	1500	5000
2400	240000	60000	1500	7500
3200	320000	80000	1500	10000
6400	640000	160000	1500	20000

Assumption: 3-5 messages per day

Figure 7.5 Relative System Paging Capacity
Source: Interviews with Paging System Operators

the pager is required. Numeric paging messages usually include the pager address, check bits and a 7 to 10 digit telephone number (approximately 25 bytes of data). Instead of a numeric phone number, alpha numeric messages may contain up to 100 characters (approximately 100 bytes). Compressed voice messages (after the removal of the pauses) as a rule last for approximately 15 seconds at 4 kbps (7.5 kbytes average). For slow speed systems, voice paging is usually performed by sending a tone alert message followed by FM modulated (non-digital) voice. Because the FM voice modulated duration is much longer than the address signaling portion, there is little difference for the capacity of system with tone/voice signaling. Figure 7.5 shows the approximate amount of paging capacity for different data rates and types of services for each paging radio channel.

A paging operator may increase their paging capacity by adding more radio channels. This means a single paging tower is likely to have several paging transmitters. Although there are many paging frequencies, many pagers receive only one frequency at a time.

Systems

Paging systems must standardize the air-interface protocols used between the antennas and the pagers. In the early days when spectrum was plentiful, no special protocol was used, a simple signal was turned on or off on a given frequency. The primary differences are in the speed of the data communication and the efficiency use of the spectrum. The following types of systems have all been standardized and accepted into the industry.

Tone Signaling

The first form of signaling to paging devices was tone signaling. Tone signaling involves the sending of several tone sequences to paging devices. The first systems sent the tones sequentially and the newer systems combined the tones. There are 2 tone, 3 tone and 5 tone systems. While many of these tone systems are still in use, most of the users have migrated to binary (digital) messaging systems.

FLEX™

The FLEX™ paging technology is the newer two-way protocol developed by Motorola to increase the capacity of paging systems. The core FLEX technology is a one-way paging system that is capable of operating at variable data rates of 1,600, 3,200, 6,400 and up to 25,600 bits per second (bps).

ReFLEX

ReFLEX is the next generation of the FLEX technology providing a reverse data channel at 9,600 bps. The protocol is designed for the most recent Narrowband PCS frequencies for two-way messaging services. ReFLEX-25 has a bandwidth of 25 kHz and a forward data rate of 12,800 bps. ReFLEX-50 has a bandwidth of 50 kHz and a forward data rate of 25,600 bps. ReFLEX service has been available from several paging carriers since 1995.

ReFLEX-25 can be set up as a simulcast system or frequency re-use system. When set up as a simulcast system, the capacity of the system is limited to the single channel that is re-broadcast by each tower. If the system approaches its maximum capacity, frequency re-use is possible to increase system capacity.

ReFLEX first sends out a broadcast locating signal and the intended pager must respond or the page/message will not be sent. When the intended pager receives the page/message, ReFLEX requires an acknowledgment indication back from the pager.

InFLEXion

InFLEXion paging systems have higher data rates than ReFLEX. InFLEXion transfers data at speeds of up to 112bps in the forward direction. These higher speeds support voice and computer applications. The voice messaging essentially downloads an electronic copy of a person's "voice mail" message to the pager. The user then "plays" back the message from the pager as if it was an answering machine. InFLEXion is also like a cellular system, in that it knows exactly which antenna is serving the pager and does not broadcast the "voice mail" to the entire system.

POCSAG

This international protocol called the Post Office Code Standard Advisory Group (POCSAG) was developed by the British Post Office in 1978. The POCSAG pager addressing format is accepted throughout most parts of the world.

The electronic address of the pager is controlled by an industry trade association (the Personal Communications Industry Association) so duplicate addresses will not be programmed into POCSAG pagers. POCSAGs support word messaging as an optional feature. The POCSAG system can operate at 512, 1,200 and 2,400 bps.

GOLAY Sequential Coding

In the 1970's, Motorola introduced a paging system called Golay Sequential Coding. The Golay format which is still used in many systems today allows numeric and text characters to be transmitted to paging devices. The Golay system uses 2 different signaling rates for addressing and data information. The addressing information is sent at 300 bps and the data information (numeric or text message) is sent at 600 bps.

European Radio MEssaging Systems (ERMES)

ERMES is an international paging technology that is used in Europe, Asia and various other parts of the world. The ERMES standard operates using 25 kHz channels and transmits at 2,400 bps. The latest generation of ERMES systems transmit at 6250 bps. For more information on ERMES, contact the European Public Paging Association (EPPA).

Radio Broadcast Data System (RBDS)

Radio Broadcast Data System (RBDS) is a low bit rate data stream that is sent along with a high power FM radio broadcast signal. RBDS is sometimes used to offer paging services. RBDS is also used to identify the radio stations call letters, to provide information about program content, and low bandwidth audio information.

When sending the paging message along with a high power radio transmission signal; the typical data signaling rates for FM Sub-band paging is 1200 bps. There are plans for higher speed radio broadcast data systems (see chapter 10).

Satellite

Satellite-based paging systems have become more popular and affordable in recent years due to the proliferation of communication satellites in orbit. Satellite systems offer unsurpassed coverage across the world. Most terrestrial based paging services only cover a certain distance outside major metropolitan areas and only a few national providers cover all metropolitan areas. The transportation (trucking) industry has come to rely heavily on satellite services because of guaranteed seamless service. Commonly satellite messaging systems can support up to 9600 bps down from the satellite to the "pager."

Services

Tone Paging

Tone paging service notifies a paging customer that a message has been sent via a tone. This tone usually is designated to mean a callback to a single location is requested. The original "beep, beep" tone pagers that started it all have become very popular as entry-level private communications systems, including restaurants (beeping servers when orders are ready) and in other centralized locations where a tone page necessitates a choice in response to only one.

Activation Fee	15
Monthly Access	5
Regional Coverage	$3.00 extra
National Coverage	$6.00 extra

Figure 7.6., Typical Tone Paging Service Rate Plan

While the popularity of these types of pagers has decreased overall, some retailers continue to offer tone pagers as "loss leaders" allowing the publication of low prices in print ads to attract attention. Still, for dispatch operators or other industries that only need a response to a central office, tone pagers offer a cost-effective solution. Typical tone-only service rates are as follows:

Numeric Paging

Numeric paging is the sending of paging messages, ordinarily telephone numbers, that are displayed on a small paging device. After the message is received, the user calls back the displayed telephone number to talk to the sender.

Numeric paging service normally involves signing up for local, regional or nationwide coverage with a monthly service charge. Both local and 800 access numbers are offered. While some carriers offer unlimited usage, many charge for additional pages beyond a pre-defined limit. A typical numeric pager service rate is as follows:

Activation Fee	$15
Monthly Access (includes 500 pages)	$7.50
Additional pages	$0.10 each
Regional Coverage	Add $3.00/mo
National Coverage	Add $6.00/mo

Figure 7.7 Typical Numeric Paging Service Rate Plan

Alpha or Message Paging

Alpha paging displays numerical and textual information. This can take many forms, from verbatim text messages to weather reports. Messages can be recorded into voice mail where operators type them up and send them out, or callers can dictate messages to live operators directly. Many carriers bundle free news, weather and sports feeds from other sources. While some carriers offer unlimited numeric usage, many charge for additional pages (both numeric and text message) beyond a pre-defined limit. Some carriers charge so much per character while others simply charge by the message. A typical alpha paging service rate is as follows:

Activation Fee	$15
Monthly Access (includes 50 pages, 80 characters each)	$25
Additional pages	$0.50 each
Regional Coverage	Add $3.00/mo
National Coverage	Add $6.00/mo

Figure 7.8 Typical Alpha Paging Service Rate Plan

Voice Paging

Voice pagers broadcast messages through a built-in speaker in the paging unit. Message volume settings can usually be set to loud, soft or private, through an earpiece. A typical voice service rate plan is as follows:

Activation Fee	$15
Monthly Access (includes 50 pages, 80 characters each)	$25
Additional pages	$0.50 each
Regional Coverage	Add $3.00/mo
National Coverage	Add $6.00/mo

Figure 7.9 Typical Voice Paging Service Rate Plan

Two-Way Paging

Two-way pagers allow fully interactive capabilities, permitting users to respond to pages by sending an original or canned alphanumeric message. Additional hardware must be purchased (or leased). Customers can lease the paging device for about $15 per month or purchase it for around $399.00. Rates for this service are as follows:

Activation Fee	$15
Monthly Access (includes 6000 characters for replies)	$35
Additional characters	$0.10 each
Regional Coverage	N/A
National Coverage	N/A

Figure 7.10 Typical Two-Way Paging Service Rate Plan

Future Enhancements

Future enhancements for paging services include broadcast services, interactive messaging and continued product miniaturization.

Broadcast Services

Broadcast services deliver a single news or other information message to large groups of receivers. AirMedia provides a broadcast service that is designed as a Wireless Internet Service for computers. This service simply requires the addition of an antenna to the computer to work. Installed on Windows® 95 or Windows® NT, the service turns your computer into a wireless receiver of "paging-like" clips of news, sports, stock prices and other valuable information from Internet content providers. This is not an online service, it is more like paging when compared to the frequency of information sought. No Internet services are required, only an agreement to connect to the AirMedia Live Internet Broadcast Network. For a very low monthly fee you can receive real-time broadcasts of information from CNN, C/NET, CBS SportsLine, Forbes, the Weather Channel and the like.

The most popular feature is enhanced e-mail alerting. Here you would be required to have an e-mail account with a service provider. The E-Mail Alert Service lets you know when you have mail. No more wasting time going online only to find that you have "No New Mail"!

Broadband Short Message Service (SMS)

Broadband wireless communications is also referred to as cellular or Personal Communications Services (PCS) and supports real-time full-duplex voice and data communications services. Narrowband (Paging) licenses do not support real-time voice communications, however Broadband PCS licenses can provide paging. Paging by digital cellular and PCS companies is done by what is known as Short Message Service (SMS). Digital cellular and PCS companies offer this SMS as paging to compete directly with the paging industry. It is two-way in the sense of guaranteed delivery, with a store and forward messaging capability. Whereas pagers are usually left on all day (and night), wireless telephone users often turn off their phone at certain times during the day. During the time a digital cellular or PCS wireless telephone is turned off, the network provider "stores" the message (page) in the network and "forwards" (delivers) the message to the unit the next time the phone is turned on.

Short message service is also used to deliver voice mail waiting indicators, similar to tone pagers that require you to call into a messaging system to receive your message. Short message service is also used to deliver information services such as sports scores and stock tickers. Coverage of short message service is the same as that of the digital cellular or PCS coverage, significantly less than a paging system's coverage.

Chapter 8

Satellite Systems

Introduction

Orbiting satellite systems provide communication services to a large geographic area. For more than 30 years, satellites have been providing voice and data communication service around the globe; however, the cost for equipment and services has been very high.

In 1997, the high cost of satellite equipment and service began to reduce dramatically. New high capacity satellites and digital technology allow for lower cost service and advanced messaging services. Early satellites were analog. After the development of digital satellites, which offer more capacity, several more satellites were put into orbit, followed by the next-generation of low orbiting satellites. These new developments are rapidly bringing the cost of equipment down by over 75%.

Although not commonly known to those in developed nations, more than half the world's population live more than two hours travel time from the closest telephone! Satellite communications are providing a way to service these remote areas with telephones, news and information feeds.

Satellites orbit in free space, where there is little or no air. In such an environment, there is little to slow the satellites down or wear them out once they are sent into orbit. The useful life time of a satellite is generally more dependent upon fuel reserves and technological obsolescence than on wear. Satellites are typically classified by the type or height of the orbit they have been placed around the earth orbit. There are three classes of satellites in orbit today: geosynchronous earth orbit (GEO), medium earth orbit (MEO) and low earth orbit (LEO). GEO satellites are positioned high above the earth (approximately at 22,300 miles) and a single satellite can cover one third of the surface of the earth. MEO satellites are commonly positioned up to 6,000 miles above the earth and a single one can cover several thousand miles. LEO systems are located at approximately 500-1,000 miles above the earth and a single one can cover a thousand miles. There are three basic types of satellite telecommunications services: broadcast international trunking, very small aperture satellite and mobile satellite service. GEO satellites are the only type of satellite that appear stationary (fixed in location) to receivers on earth compared to MEO and LEO satellites that regularly move across the horizon. There are three key portions to satellite systems: satellite section, ground section and end user equipment.

The specific frequencies dedicated to satellite services are in the microwave bands (such as the C band, KU, KA and L bands). The FCC and other global entities control these frequencies. Because satellite transmission covers very large geographic areas that may extend outside a country, there is a key issue of interference control. There are so many satellites that they have to maintain a minimum of two degree separation between them.

Satellite Segment

A satellite is a space vehicle that orbits the earth, and which contains one or more radio transponders that receive and retransmit signals to and from the earth. The size and weight of satellites varies from 1 to over 20 meters in length and 90 lbs (20 kg) to over 8800 lbs (4000 kg). There are several 4000 kg GEO satellites being prepared for launch in 1998. Satellites contain a power supply, position control system, transmitters and receivers (called transponders), and an antenna system.

The power supply for a satellite ordinarily consists of solar panels and a backup battery. The amount of power used by satellites varies from a few hundred watts for small low earth orbit satellites to several thousand watts for large high earth orbit satellites.

Satellites do not automatically stay in their desired location, because of gravity effects from the irregular earth, sun, moon, and other planets. The position of the satellites must be continuously monitored and adjusted (an activity called "station keeping"). The ability to control the stability (wobble) of a satellite is controlled by having a section of the satellite spin at 50 to 100 revolutions per minute. The speed or amount of spin is able to control the angle or relative position of the satellite. The general altitude is normally controlled by a pressurized gas system (generally hydrazine). This means that the life of the satellite system is usually determined by the amount of hydrazine it can carry.

For example, GEO satellites are kept in their correct latitude and longitude by rocket motors that oppose the gravity effects, so they appear to remain stationary with respect to the earth, and ground antennas do not have to track them. The amount of fuel available to do these station-keeping maneuvers generally determines the life of the satellite. Some are allowed to drift north and south and still operate after nearly running out of fuel, because the North-South station keeping uses a 100 times as much fuel as the East-West station-keeping. This requires ground antennas to track them in most cases.

Most satellites in existence today are three-axis stabilized. The ability to control the attitude stability (rotation) of these satellites is normally controlled by momentum wheels, magnetic torquers, and sometimes small rocket motors for three-axis stabilized satellites. Spin-stabilized satellites, of which there are relatively few in the GEO type, are stabilized by the gyroscope effect of spinning. The payloads are stabilized by electric motors and earth sensors.

The main purpose of satellites is to receive radio signals from earth and retransmit these signals back to earth. Usually, this is accomplished by a transponder. Such a transponder is called "bent-pipe." The transponder receives a signal on one frequency, converts the frequency, amplifies the signal and then sends the signal back to earth. There may be 40 or more transponders on a single satellite, each having a radio channel bandwidth of 80 MHz (or more) Some modern satellites today do on-board digital processing of the signals. These satellites are much more complex and expensive than the simple transponder types described above.

The antenna system of a satellite usually consists of several directional antennas, allowing the satellite to direct its radio energy to specific locations. The radio coverage area provided by a satellite is called its footprint. The footprint of a single satellite can be thousands of miles in diameter, which may be enough to cover an entire continent. Sometimes, however, spot beams are used that cover only a hundred miles or so in order to concentrate signals into a small area and develop stronger signals because of higher antenna gain.

Some systems have multiple satellites (such as MEO and LEO) and these systems often have spare satellites in orbit in the event of an equipment failure. Some GEO systems, such as Intelsat, also orbit spare satellites for both equipment failure and to pick up increased demand.

Ground Segment

The ground segment of a communication satellite system contains the gateways that send and receive information signals from the satellite, switching or routing facilities and a satellite control center.

The gateways provide access to the space segment and interface to public and private data networks. The major elements of a gateway include gateway earth stations, each of which is composed of an antenna, controller and radio equipment. The systems used in a gateway typically have redundant assemblies to allow rapid restoration of service in the event of failure. For GEO systems that have only one satellite, there may be only one gateway. For MEO and LEO systems, gateways are located throughout the world. These facilities monitor and manage all network elements to ensure continuous, consistent operations in the provision of quality service.

Gateways connect with a media source or switching system that provides interconnection to the terrestrial networks. A media source may consist of television channels that come from a nearby studio or via a network information feed. Switching systems for mobile satellite systems (MSS) are used to connect and process communication paths.

All satellite systems have some form of a satellite control center. The satellite control center monitors and controls the position of the satellite, sends commands for various satellite functions, and monitors the telemetry regarding satellite health and condition. Many of the control center antennas autotrack and range on the satellites for determining their position.

Subscriber Equipment Segment

End user equipment (typically called subscriber units) consists of an antenna, radio receiver, transmitter (only for two-way systems) and an interface converter (such as a video or audio interface) depending on the application.

There are various types of subscriber units, some of which are intended for general use, and some of which are designed to support specific applications. Subscriber units that are used for mobile or portable voice communication may be capable of several services such as voice, messaging and data. They usually appear very similar to a cellular telephone. Other devices, such as digital television receivers or meter reading devices, are designed specifically for their application.

Broadcast

Broadcast service allows the same signal, such as television or audio signals, to be sent to all the receivers. Satellites used for broadcast services are commonly GEO, allowing fixed antennas dishes on the earth to stay pointing directly at them. These types of satellites are most familiar because they have been around more than 30 years.

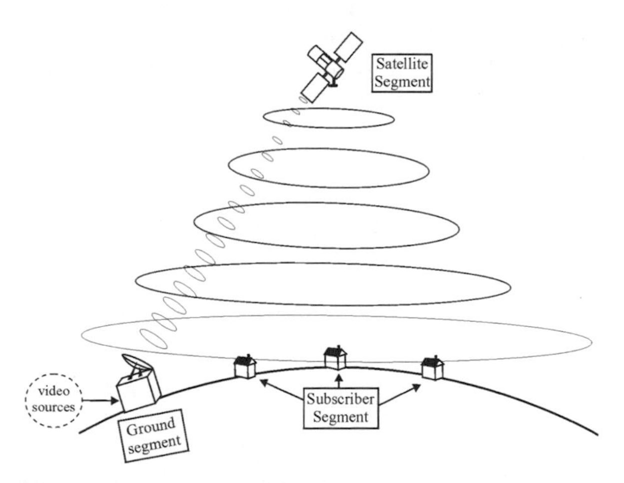

Figure 8.1 Basic Broadcast Satellite System

Figure 8.1 shows a basic broadcast satellite system. In this diagram, several video signals are combined onto a single radio channel that is uplinked to the satellite. The satellite transponder converts the uplink signal and retransmits it back to earth so that homes with satellite receivers can view the television channels.

In a typical broadcast satellite system, there are several program sources (television signals) that are combined into one radio transmission signal. This signal is sent up to a satellite (on the uplink) where it will be re-transmitted back down (the downlink) to a large geographic area.

The radio frequency broadcast signal travels great distances from GEO satellites, so special antennas (dishes) are used to gather enough radio signal energy to receive the broadcast signal. The television, news and entertainment industries utilize these types of satellites to distribute and receive video and audio programs.

One of the latest advances in broadcast satellites is direct broadcast satellite (DBS) service. DBS service uses high power (over 100 watts per transponder) satellites and compressed digital signals to allow a single small satellite dish (ordinarily 18 inches in diameter) to receive over 150 channels of programming. DBS receivers have become one of the fastest growing consumer electronic products with growth rising from under 300,000 units in service in 1995 to over 6 million units in service in 1998.

Digital audio broadcasting (DAB) is another broadcast system that can use satellite systems to deliver high quality digital audio to a large geographic area. Systems are being tested around the world to deliver DAB from satellites as well as from land based (terrestrial) antennas.

Very Small Aperture Terminals (VSAT)

Very Small Aperture Terminals (VSAT) are ground-based satellite terminals that offer two-way data access to satellite systems. Typical uses for VSAT networks are the broad distribution of data to many receivers. An example of this service is the distribution of price or inventory levels to a chain of stores throughout a large geographic area. They are also used for each store to send inventory and sales information back to headquarters. An important feature of a VSAT service is ease of deployment; installation takes approximately two hours. Companies are now installing VSATs at the rate of more than 1,500 per month.

Because VSAT systems can often communicate in two directions, fast-response protocols are used for time-sensitive transactions such as credit card purchases and hotel reservations. When data is being sent through a VSAT network, it uses efficient data compression for file transfers.

An example of a use for VSAT systems is the sending of new pricing information from a corporate headquarters to its stores located throughout the country. The corporate headquarters provides a data signal to the satellite gateway which adds the companies' addresses to the message and inserts the data into the uplink satellite data channel. The satellite retransmits the data and the stores which have a VSAT receiver and the correct address code, will receive it. Some stores (such as Walmart) may have over 2,000 stores that simultaneously receive the message.

Mobile Satellite Systems

Mobile Satellite Systems (MSS) provide two-way communication with mobile satellite radios. MSS systems can be of the GEO, MEO or LEO type. MSS can be divided into maritime mobile satellite services (MMSS), land mobile satellite service (LMSS) and aeronautical mobile satellite services (AMSS).

An example of a mobile satellite system is communication between a ship and a landline telephone system. To establish communication, a shipboard satellite telephone first dials the telephone number of a landline telephone. The satellite radio then sends a request to the satellite for service and the dialed digits are retransmitted (relayed) to the satellite gateway where it can be connected to the public switched telephone network (PSTN). If the service is authorized, the gateway will dial the office telephone and when the call is answered, the audio from the shipboard satellite telephone will be connected to the audio of the office telephone (through the phone system).

Satellite telephones have been relatively large (briefcase size) until recently. Due to the distance of GEO satellites, mobile satellite phones had to use high power amplifiers and relatively large directional antennas. With the introduction of low earth orbit systems, it is possible to use handheld portable satellite telephones. The MEO, and some new GEO systems with very large satellite antennas, will also support handheld portable phones.

Many mobile satellites allow dual mode satellite and terrestrial (ordinarily cellular) service. This service allows the mobile telephone to use the cellular system if it is available. If the cellular system is not available, the mobile telephone will access the satellite system.

Market Growth

In the United States, there were approximately 130,000 mobile satellite telephones in use during 1998. As a result of the introduction of new low cost satellite technology, it is predicted that there will be over 8 million satellite telephone and paging devices in use in the United States by the year 2008. Estimates predict that by 2010, there could be as many as 40 million mobile satellite subscribers globally. Maritime satellite applications alone are growing by more than 25% annually. Aeronautical applications will also contribute to the growth with a projected 3,500 aircraft offering satellite telephones by the year 2000.

According to brokers, ABN Amro Hoare Govett, the world's mobile telecommunication market is worth approximately $3.2 billion. Estimates indicate the mobile satellite market could be worth around $500 million in 2000, increasing to more than two billion dollars by the year 2005.

Some estimates predict that over 1,000 Low Earth Orbit/Medium Earth Orbit satellites will begin service in the next 10 years. Almost one-half of these will be for broadband multimedia communications services. The remaining will be for both voice and data services as well as data only services.

Including home satellite televisions systems, the industry generated $11.32 billion in sales for satellite receivers, representing 8.5 million units sold in 1996. Regarding just the direct-to-home satellite systems in the U.S., the growth has greatly impacted the cable industry. Though satellite service was available to the public in 1987, it was not until 1994 when the smaller, compact satellite dishes became available that direct broadcast television receiver sales began their rapid climb. By 1996, a total of 10 million (or 10%) of the households in the U.S. had converted to satellite TV service.

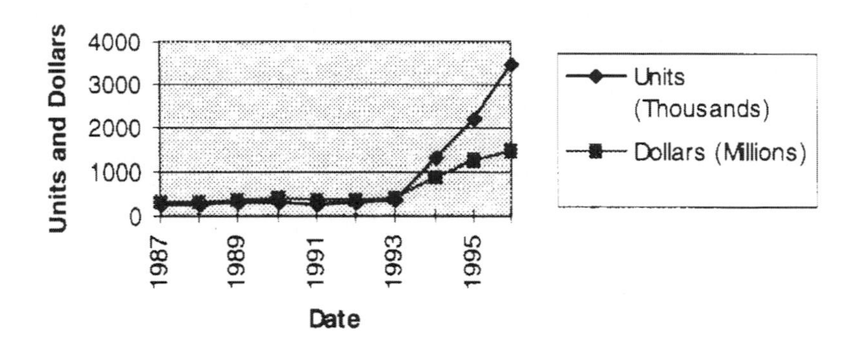

Figure 8.2.,US Direct-to-Home Satellite Sales
Source: EIA Electronic Market Data Book

Figure 8.2 demonstrates the annual sales increase in the growth in the US satellite television industry. As a result of this growth, 82% of the consumers downgraded their cable subscriptions, with 64% of those eliminating their cable service all together. This resulted in an estimated $1.4 billion annual loss to the cable industry. This growth is estimated to continue to accelerate due to the small size of the dishes, product satisfaction, and ease of installation.

As more satellites for commercial communications are launched, the market is entering an unparalleled stage of growth. With the various constellations of LEO satellites providing global mobile (portable) telephone communication services, communications satellites will experience the most growth.

Consolidation among companies and countries is also beginning to shape marketing trends. Higher costs as well as risk sharing are all contributing to the alliances.

Figure 8.3 shows the projected US growth of the mobile telephone satellite market. Not included in this chart are Digital Broadcast

	1998	1999	2000	2001	2002	2003	2004	2005	2006	2007
Voice	120	180	230	310	470	680	1160	2080	3650	5520
Data	90	120	180	240	275	365	425	830	1140	2010

Figure 8.3 Satellite System Market Growth
Source: APDG Research

Satellite (satellite television) or VSAT service. The market is divided into voice customers and paging/messaging customers. Voice customers may be using a combination of cellular and satellite telephones. Initial customers are likely to be companies or agencies that require large, reliable geographic coverage. This would include government agencies, trucking companies and possibly environmental monitoring companies providing service in rural areas. Many companies are experiencing launch delays due to El Nino effects and even accidents, which may contribute to the delay of overall market growth.

Technologies

Geosynchronous Earth Orbiting (GEO) Satellites

Geosynchronous earth orbit (GEO) satellites are located approximately 22,300 miles above the surface of the earth. The reason for choosing this particular height above the earth is the precise balance between the gravity pull to the earth and the centrifugal force (speed) of the satellite spinning around the earth, thus the satellite spins with the earth's rotation. It is at this height where the satellite will appear to be fixed in position (stationary) above the Earth because it is spinning at the same rate as the earth (1 revolution per day). In 1963 Hughes Aircraft and NASA achieved the first geosynchronous orbiting satellite.

The primary advantage of a GEO satellite system is that only one satellite is required to provide radio coverage to one-third of the earth's surface. GEO satellites also have an average life span of 10 - 15 years.

Because GEO satellites are located a long distance away, radio transmissions through these satellites have an approximate delay of 1/2 second (1/4 second each way). This transmission delay can result in difficulty when thc system is used for voice services.

GEO satellites have a relatively simple design but are expensive to build due to their large physical size, which is necessary to provide the high power signal to receivers on earth. Historically, Inmarsat mobile satellite telephones that communicate with GEO satellites have high

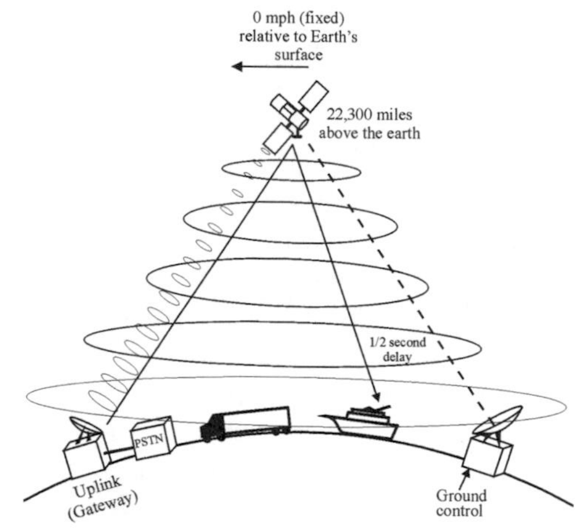

Figure 8.4 GEO Satellite System

power transmitters and directional antennas to send the signal the long distance to the satellite. Briefcase size antennas are required along with substantial power requirements. There are now several GEO satellite systems under design with very large satellite antennas (10 to 14 meters, 33 to 46 feet diameter) that will communicate with hand-held cellular-like telephones.

Figure 8.4 shows a GEO satellite system. This diagram shows that the GEO satellite is connecting a telephone caller on land with a ship. The telephone call is routed through the gateway to the satellite. The satellite transponder converts the frequency and retransmits the signal back to earth. A satellite telephone on the ship receives the radio signal and converts it back to the original audio signal. The diagram also shows that a separate ground control facility is used to monitor and control the position of the satellite.

Medium Earth Orbiting (MEO) Satellites

Medium earth orbit (MEO) satellites are positioned between 1,000 to 6,000 miles above the earth. Unlike GEO satellites which appear stationary above the earth, MEO satellites move slowly through the sky as seen from an observer on the earth. A typical MEO satellite will circle the earth in approximately 6 hours. Among other benefits, this provides the potential for global coverage if the countries which the satellites move over allow radio transmission into their country.

Because the MEO satellites are located closer to the earth, the radio coverage footprint from MEO satellites is somewhat smaller than from GEO satellites. This means 2 to 3 MEO satellites are required to cover the same area as a GEO satellite. Because these satellites move, several MEO satellites must be positioned around the world to allow continuous coverage of a specific area. As one satellite moves out of the radio coverage area, another satellite can take its place. With approximately 10 to 12 satellites, global radio coverage is possible.

Because they are much closer to the earth than GEO satellites, MEO satellites only have about a 1/10th second or less delay time for transmitting a voice/data signal. They are also sometimes of a smaller size than the GEO satellites and therefore they are less expense to build and launch. Similar to GEO satellites, MEO satellites have an average life span of 10 to 12 years. MEO satellites do allow mobile communications to low to medium power handheld portable satellite telephones.

Figure 8.5 shows a MEO satellite system. In this diagram, several satellites circle the earth at several thousand miles per hour. In this example, a landline telephone is communicating with a portable satellite telephone. The telephone call is routed through the gateway to the satellite. The satellite transponder converts the frequency and retransmits the signal back to earth. The portable satellite telephone receives the radio signal and converts it back to the original audio signal.

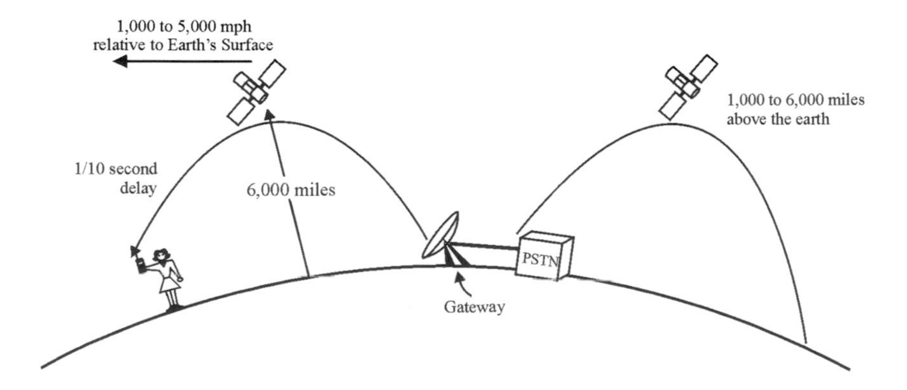

Figure 8.5 MEO Satellite System

Low Earth Orbiting (LEO) Satellites

LEO satellites are located approximately 450-1,000 miles above the Earth. Because LEO satellites are located so close to the Earth's surface, LEO satellite systems normally provide mobile satellite services (MSS) to handheld or mobile satellite telephones.

Also because LEO satellites are located very close to the Earth, each satellite must move at approximately 17,000 miles per hour to avoid falling into the Earth. LEOs circle the earth in approximately 90 minutes. LEO satellites are technically much simpler and more robust than geosynchronous satellites and are less likely to suffer catastrophic failure during deployment or during the satellite lifetime.

The radio coverage footprint of a LEO satellite is small, so 50 or more satellites are required to achieve continuous coverage. As one LEO satellite moves out of the radio coverage area, another satellite can take its place.

Because LEO satellites are only 450 to 1000 miles above the earth (10 to 50 times closer to the earth than MEO or GEO satellites), LEO satellites have a negligible delay time for voice/data signals. LEO satellites are also much smaller and have a much lower cost to build and launch than other types of satellites. They have an average life span of 5 to 8 years, primarily because of radiation effects. LEO satellites do allow mobile communications to low power handheld portable satellite telephones which do not need highly directional antennas. In practice, the transmit power level can be lower than 1 Watt for

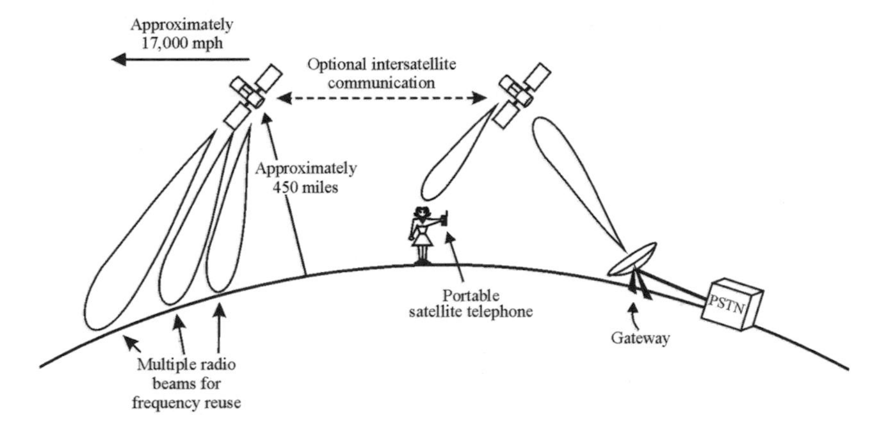

Figure 8.6 LEO Satellite System

portable LEO satellite terminal equipment.

Figure 8.6 shows a LEO satellite system. In this diagram, a portable satellite telephone is communicating with a landline telephone. The satellite telephone communicates with the closest LEO satellite. Because LEO satellites fly very close to the surface of the earth, they go across the visible horizon in approximately 10 minutes in reference to a mobile satellite customer's location. When the first satellite moves out to the horizon, another LEO satellite becomes available to continue the call. However, robust network communications need to be in place to maintain calls (especially data transmission) within this period. Some systems will use satellite diversity to allow talking through more than one satellite at a time, avoiding call "dropouts" from signal blockage.

Global Positioning System (GPS)

Global Positioning System is a navigation system that uses satellites to act as reference points for the calculation of navigational position. GPS is used extensively by the military and aircraft GPS chipsets are now being incorporated into wireless devices, including phones, personal digital assistants (PDA's), as well as automotive applications.

In the late 1980's, the military began to allow GPS technology to be used for public use. Due to the much larger number of units in production, the price for GPS equipment was dramatically reduced. Because the satellites are maintained by the government, there is no cost to use GPS service.

There are several systems used for global positioning. These include the Russian Global Navigation Satellite System ("Glonass"), the United States global positioning system (GPS) and another United States positioning system called "Sat-Nav".

The GPS system is based on 24 orbiting satellites called "NAVSTAR." GPS receivers can calculate their exact location (height and location (Longitude and Latitude)) by measuring the distance from four or more satellites. Consumer devices are accurate within one city block, and commercial/military devices can measure to the millimeter (less than 1/16 inch). With the addition of software grid information, a real-time driving location map can be generated and available to those people "who never get lost".

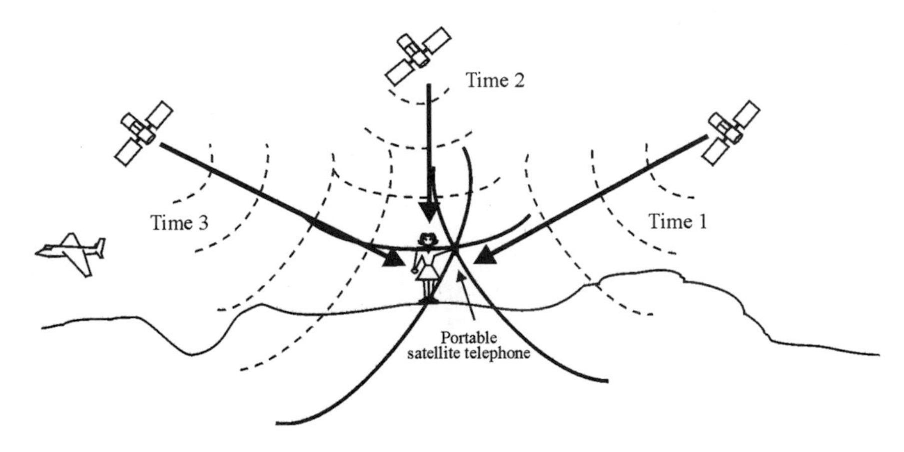

Figure 8.7 Global Positioning System

Figure 8.7 shows a global positioning satellite (GPS) system. This system uses three or four satellites to triangulate the GPS receiver's position. Each satellite transmits its exact location along with a timed reference signal. The GPS receiver can use these signals to determine its distance from each of the satellites. Once the position and distance of each satellite is known, the GPS receiver can calculate the position where all these distances cross at the same point. This is the location. This information can be displayed in latitude and longitude form or a computer device can use this information to display the position on a map on a computer display.

Spatial Division Multiple Access (SDMA)

Spatial division multiple access (SDMA) is a system access technology that allows a single transmitter location to provide multiple communication channels by dividing the radio coverage into focused radio beams that reuse the same frequency. To allow multiple access, each mobile radio is assigned to a focused radio beam. These radio beams may dynamically change with the location of the mobile radio.

Satellite systems use spatial division multiple access (SDMA) to increase their capacity. Each satellite has multiple antennas, or a phased array with multiple beams, which provide radio coverage to 10 to over 100 smaller geographic coverage areas within their satellite radio coverage area.

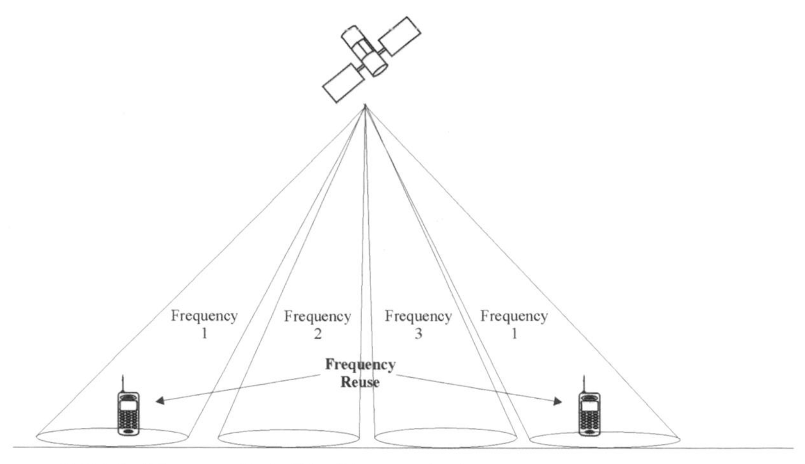

Figure 8.8 Mulitbeam Satellite Transmission

Because a single satellite can provide coverage to an area several thousand miles wide, SDMA technology can use that single satellite to provide several radio beams, some of which reuse the same frequencies. This increases the number of simultaneous channels a single satellite can serve.

Figure 8.8 shows a satellite system that uses SDMA technology. In this example, a single satellite contains several directional antennas. Some of these antennas use the same frequency. This allows a single satellite to simultaneously communicate to two different satellite receivers that operate on the same frequency. Usually beams that are separated by more than two or three half-power beamwidths can use the same frequencies, as shown in the figure.

Multiple Access Technologies

The radio communication systems used by satellite have changed over the years. Some satellite radio channels continue to use analog modulation, however most satellite systems are converting to digital transmission. The method used to share digital channels varies depending on the system and equipment that will be used to access the satellite system. The different types of access technology used for communication satellites include time division multiple access (TDMA), frequency division multiple access (FDMA), code division multiple access (CDMA), and time division duplex (TDD). The access technology types can vary between the uplink and downlink channels for satellites that do on-board processing.

Another system that is used to increase the efficiency of satellites is digital speech interpolation (DSI). DSI is a technique that dynamically allocates time slots for voice or data transmission to a user only when they have voice or data activity. This increases the system capacity as transmission for other users can occur when others are silent.

New Systems

The FCC requested proposals for new satellite systems in two new identified frequency bands in 1997. They are: 1) 1990-2025 MHz U.S., 1980-2025 MHz elsewhere for uplink, and 2165-2200 U.S., 2160-2200 elsewhere for the downlink, called the 2 GHz band. 2) 48.2-49.2 GHz uplink, and 37.5-38.5 GHz downlink, called the 38/49 GHz band. They received 9 applications for the 2 GHz band, and 16 for the 38/49 GHz band. The frequencies are not yet allocated, but these activities indicate the trends in the next few years for satellite communications. Bidders proposed both voice and data services.

Mobile Satellite Systems

There are hundreds of satellite systems that are in commercial use and several more that are in final testing or nearing commercial deployment. The following list includes many of the satellite systems that are in use.

INMARSAT

INMARSAT, the International Maritime Satellite Organization, formed in 1979, is now backed by the governments of 75 member countries. Its first satellite (INMARSAT-A) became operational in 1982, supporting voice and low-rate data applications with analog FM technology. In 1993, 30,000 ground terminals were in operation. The next generation of INMARSAT satellites (INMARSAT-B and C) used digital technology. With the introduction of INMARSAT-M in 1996 it is now possible to use laptop computer-sized satellite terminals for voice and low-rate (2.4 kbps) data transmission. INMARSAT transmits in the 1.5 GHz range which is allocated worldwide for satellite transmissions. Inmarsat has formed a new company, ICO, which will operate a family of MEO satellites for cellular-like communications.

OMNITRACS

In the late 1980s, QUALCOMM deployed the OMNITracs vehicle-tracking and communications system for both North America and Europe. The service provides two-way messaging and automatic position reporting. By 1997 more than 200,000 trucks, most of them in the United States, were equipped with the system.

American Mobile Satellite Corporation (AMSC)

American Mobile Satellite Corporation (AMSC) is the first satellite-based mobile voice and data communications service in the United States being offered to the general public. As of 1998, AMSC had over 85,000 customers [1]. AMSC offers voice and data services, using its own dedicated geosynchronous satellite. As of May 1998, AMSC has taken its satellite out of U.S. service and is using its Canadian dual satellite MSAT to provide U.S. service. MSAT is identical to the AMSC satellite. AMSC offers a wide range of public services to land mobile users, maritime, aeronautical, and fixed-site applications. A major focus is on land-mobile cars and trucks. The mobile portable terminals are briefcase size, whereas this system mounted in vehicles appears normal.

In 1998, AMSC purchased ARDIS, a terrestrial packet data radio service to allow the combination of satellite and land based data service.

COMSAT

COMSAT was formed by an act of Congress in 1962 and represented U.S. commercial interests in satellite technology development. As an international, government-chartered organization, COMSAT was tasked to coordinate worldwide satellite communications issues. Comsat is the U.S. Signatory for Intelsat, and provides mobile services by virtue of being the U.S. Signatory for Inmarsat.

ORBCOMM

Orbcomm is the world's first commercial LEO satellite data communications and position determination system. Orbcomm provided limited commercial service in 1996. Orbcomm requires 1,000 times less power than the traditional geosynchronous satellites. The basic system includes 26 satellites in the constellation transmitting in the 137-150 MHz range. An additional eight satellites will be added to provide for increased capacity and improved service in 1999. The ORBCOMM system is primarily designed to provide data services.

The Orbcomm system uses less complex and lightweight satellites (which weigh approximately 90 pounds each). The height of the satellites above the earth varies from 740 to 825 kilometers so the speed that they travel also varies. The Orbcomm satellites are unique as they are designed to fly and can be adjusted in height similar to how an airplane adjusts its height. The Orbcomm satellites are smart satellites which use an on-board computer system that controls the radio channel assignment dynamically to avoid interference.

Iridium

Iridium, which began in 1990, is a $4 billion LEO project designed primarily for world-wide cellular telephone coverage. The Iridium system has 66 satellites that are used to provide global radio coverage.

Each Iridium satellite projects a grid of beams (cells) onto the Earth's surface in a manner similar to, but larger than, the grid of cell sites in a cellular telephone network. This allows for frequency reuse and much higher system capacity.

A unique attribute of the Iridium system is that each satellite has the ability to directly communicate with adjacent satellites. This permits calls to be connected from satellite to satellite without the assistance of a ground station. The Iridium system has the capability of switching calls directly through the satellite network.

The digital voice signal is at 4800 bps and the data channels are typically at 2,400 bps. The Iridium system is capable of providing voice, data, paging and wireless fax services. The first Iridium satellites were launched in 1997. Initial tests have shown that the Iridium system works well. Iridium is scheduled to be in full operation by the third quarter of 1998.

Globalstar

Globalstar is a joint venture between Qualcomm and Loral Corporation consisting of 48 LEO satellites at an altitude of 875 miles. The Globalstar system is expected to offer voice, paging and messaging services throughout the world. In addition to the basic voice services, it is expected that Qualcomm's OmniTRACS position location and messaging system will be deployed on the Globalstar System.

By the end of 1998 it is expected that 44 Globalstar satellites will be orbiting. The remaining 12 (48 operational plus 8 spares) will be launched early in 1999. The first Globalstar satellites were launched in 1998. Initially, Globalstar will deploy only 24 satellites. This will be increased to 48 satellites. The system has the basic data rate of 9.6 kbps for voice communication and up to 7.2 kbps for data. The access technology for the Globalstar system is CDMA. This is similar to the CDMA system used for cellular communications. Various dual-mode and tri-mode hand-held telephones will be used to provide combined cellular and satellite service.

Globalstar satellites use a simple "bent pipe" approach which simply amplifies, upconverts or downconverts, and reflects the radio signals back to Earth. This approach smaller, less complex satellites with all the call processing and switching operations performed on the ground. Each satellite has 16 directional antennas to create spot beams. This increases the system capacity and better controls the radio coverage area.

ICO Global Communication

ICO is a company that was initiated by Inmarsat to be a provider of mobile satellite services. The ICO satellite system will consist of 10 MEO satellites at an altitude of 10,355 km in two planes 90° apart.

Orbit inclination is 45°. It will have 12 Gateway sites operating at C-band, 5 and 7 GHz. It will service users with a TDMA format to provide voice service.

Teledesic

Teledesic is a "Big LEO" satellite system that is proposed to provide high speed data services (broadband) to consumers. The term "Big LEO" is used to indicate in excess of 200 satellites. The Teledesic system is a $9 billion project initiated by Bill Gates and Craig McCaw. Teledesic is planned to become available in the year 2001.

The Teledesic system is designed to transmit data from 16 Kbps to 2 Mbps inexpensively with high quality and low delay. This will allow Teledesic to provide for both voice and data services. Teledesic is primarily expected to provide affordable high speed Internet access to remote areas. Teledesic has planned for 288 satellites in 21 different levels (orbital planes). At any time, each Teledesic subscriber unit will be able to connect to at least two of them from anywhere in the world. Teledesic satellites have the capability to communicate with adjacent satellites at speeds of 155 Mbps. Teledesic has recently formed business arrangements with Boeing and Motorola for design and construction of the satellites. (Teledesic changed from 840 to 288 satellites shortly after joining forces with Boeing.)

Services

Satellite services include long term leasing of satellite channels, wide area dispatch services, voice and data services. The pricing of satellite service is based on many variables, including the availability and cost of substitute services, the cost of providing service and the nature of the customer application. Pricing generally is based on a wholesale pricing structure that incorporates an initial activation charge, a recurring monthly charge for access to the system and charges based on the customer's usage. Additional pricing considerations include access priority and messaging transport usage.

As competition increases in the satellite industry, it is likely that multiple pricing alternatives will be offered including peak/off-peak, volume discounts and annual contract commitment options. Because the satellite industry uses many resellers of service, the usage pricing for services will be largely outside the control of the satellite system owner and will be established by value added resellers (VARs).

Transponder Leasing

It is common that a portion of the transponder such as a single radio channel or portions of a transponder's radio channel may be leased out for the life of the satellite to a value added reseller (VAR). The VAR commonly sub-divides the transponder channel to provide temporary communication services to their clients which include video links for news services, teleconferencing, virtual local area networks (VLANS), VSAT, position tracking service, global voice and messaging communications.

The typical lease for a transponder can be several hundred thousand dollars per month depending on the capacity of the channel (e.g. 6 MHz), coverage area and other system services that are provided by the satellite owner.

Dispatch

Satellite systems allow for cost effective wide area (typically nationwide) dispatch services. Dispatch services allow a single user to communicate with one to many users simultaneously in a customer-defined group using a push-to-talk device. Dispatch service can be used to coordinate operations between workers that are operating over wide and/or remote areas.

Companies that benefit from dispatch services include oil and gas pipeline companies, utilities and telecommunications maintenance fleets, state and local public safety organizations and public service organizations with a requirement to communicate on a nationwide basis.

The service fees for dispatch type services vary from a fixed monthly fee with an almost unlimited amount of service to per minute charges ranging up to $7.50 per minute.

Voice

Satellite telephones are usually capable of two-way voice, facsimile and data services. In the early 1990's, the hardware cost for a mobile satellite telephones ranged from approximately $4,000 to $10,000 each. With the introduction of LEO satellite systems that can use low power transmitters, handheld satellite telephones are expected to cost $750 or below [2].

Users of satellite telephone voice services are usually charged a fixed monthly access fee and variable usage fees. Monthly bills for satellite voice customers range from under $50 per month for occasional users to over $150 per month for moderate usage maritime users [3]. The rates for usage are expected to drop as new competitors enter the marketplace.

Figure 8.9 shows a expected voice satellite rate plans for different types of satellite system. Per-minute charges for telephone service are expected to range from below one dollar to several dollars per minute.

	Globalstar	Iridium	ICO	Odyssey	Ellipso	Constellation
$/Minute	$0.35-$0.55	3	$1.00-$2.00	< $1.00	< $0.50	< $0.50
Terminal Cost	750	3000	$1000-$1500	$300-$500	N/A	N/A
System Cost	$2.5 Billion	$3.5 Billion	$2.6 Billion	$3.2 Billion	$0.9 Billion	$1.2 Billion
Satellite Life (years)	7.5	5	10	12	5	5
Digital Interface	CDMA	TDMA	TDMA	OCDMA	CDMA	CDMA
Orbit	LEO	LEO	MEO	MEO	LEO	LEO
# Of Satellites	48	66	12	12	16	46
In-Service Date	1998	1998	2000	2001	2000	2000

Figure 8.9, Projected Mobile Satellite Rate Plan
Source: Paine Webber

Data

Satellite data service users are usually charged a monthly access fee that includes a fixed increment of data usage that may consist of vehicle location reports or data messages. Usage beyond the fixed increment is metered and charged on a variable basis depending on the length and mode of transmissions. Satellite data services offer a wide variety of volume packaging and discounts, consistent with the demands of the targeted markets.

Basic data services include fixed and mobile asset tracking, monitoring and messaging services. Primary fixed asset monitoring applications include electric utility meters, oil and gas storage tanks and wells, oil and gas pipelines and environmental monitoring. Mobile asset tracking includes tracking the location and status of commercial vehicles, trailers, shipping containers, rail cars, heavy equipment, fishing vessels and barges and government assets. Messaging services include short alphanumeric paging-like communications services and wireless email.

The cost for data transmission via satellites has dropped from approximately 1 cent per byte of information ($10 per kilobyte) to under 0.1 cent per byte ($1 per kilobyte). It is likely the cost per kilobyte will drop to below 50 cents per kilobyte by the year 2000 [4].

Future Enhancements

Satellite technology is changing rapidly. Likely future advances will include high speed Internet access and airborne satellites.

High Speed Internet

Satellite technology is evolving to allow higher speed digital services through the use of LEO systems, SDMA technology and digital access technology. Because LEO systems have many satellites, their ability to reuse frequencies often increases the total data rate capacity of the system. Because each of the new LEO satellites uses multi-beam antennas that permit frequency reuse, this also increases the total data capacity of the system. More efficient digital transmission technologies (more bits per Hertz) are continuing to be developed. The combination of these key technologies and others will allow user data rates to approach 2 Mbps.

Airborne Communication Platforms

Another future satellite technology is the use of airborne sky stations. These airborne platforms will have an antenna the size of a football field that will be located just miles above the surface of the Earth. With the large antenna size and small distance from the surface of the Earth, radio signals transmitted from low power handheld units can be handled efficiently.

One company, Sky Station International, has filed applications with the Federal Communications Commission proposing this new global wireless communications system. Using a network of 250 stratospheric balloons (dirigibles), Sky Station plans to offer wireless services to more than 80 percent of the world's population by 2002. Each lighter-than-air dirigible will comprise two blimp-like aircraft to suspend the large antenna between them. Once completed, the network (which reportedly costs $4.2 billion) will provide 64-Kbps digital broadband service to 1.5 billion customers worldwide. The network will provide services such as wireless Web access and picture-phone connectivity. Sky Station has asked the FCC to designate frequencies at 47 GHz for the system's operation in the U.S.

1. "AMSC 10K report", United States Securities and Exchange Commission, March 31, 1998.
2. "Globalstar 10k report", United States Securities and Exchange Commission, March 31, 1998.
3. "AMSC 10K report", United States Securities and Exchange Commission, March 31, 1998.

Chapter 9
Fixed Wireless

Introduction

Fixed wireless is the use of wireless technology to provide voice, data, or video service to fixed locations. There are several fixed wireless systems that can replace or bypass services that have traditionally been provided by copper wire or fiber cable. Wired systems that may be replaced or bypassed include wired telephone service, high speed telephone communication links, cable television systems and local area network systems.

When the Telecommunications Act of 1996 was passed by Congress, the local telephone monopoly in the United States was expected to end, just as it did in the United Kingdom in early 1990s. Like the UK and the US, many countries are now deregulating telecommunications (PTT monopolies) by allowing new competition for local telephone, long distance telephone and cable services. For example, India and Indonesia will soon have wireless local loop providers competing with the existing telephone company.

The terminology of "local telephone service" and "cable service" as we know it is changing. Already some cable companies are offering local telephone service and some telephone companies are offering television services. New companies are being formed to compete with the local telephone and local cable companies. Many of these new entrants will use wireless as their access to the local loop. The use of wireless allows rapid deployment of services and reduces the cost of installing cables to each residence or building.

Using wireless systems instead of wired systems allows new entrants to keep the system construction costs down while deploying the systems quickly (in months instead of years in some cases). The basic fixed wireless technologies that are being introduced to replace or bypass cables include wireless local loop (WLL), wireless cable, wireless bypass, and wireless local area networks (WLAN). These fixed wireless services can provide local dialtone voice service, high speed data and video service. In some cases, a single fixed wireless system may provide all these services at the same time.

Wireless Local Loop

Wireless local loop (WLL) service refers to the distribution of telephone service from the nearest telephone central office to individual customers via a wireless link. In some cases, it is referred to as "the last mile" in a telephone network. This term is a bit misleading, though, because the coverage area of a WLL system may extend many miles from the central office.

Local exchange carriers (CLEC) are competitors to the incumbent local exchange carriers (ILECS) and are likely to use WLL systems to rapidly deploy competing systems. If CLECs do not use wireless systems, they must either pay the existing phone company for access to the local loop (resale) or dig and install their own wire to the local customers. Many countries, that do not have large wired networks such as the United States, are using wireless local loop as their primary phone system.

Figure 9.1 shows a wireless local loop system. In this diagram, a central office switch is connected via a fiberoptic cable to radio transmitters located in a residential neighborhood. Each house that desires to have dialtone service from the WLL service provider has a radio receiver mounted outside with a dialtone converter box. The dialtone converter box changes the radio signal into the dialtone that can be used in standard telephone devices such as answering machines and fax machines. It is also possible for the customer to have one or more

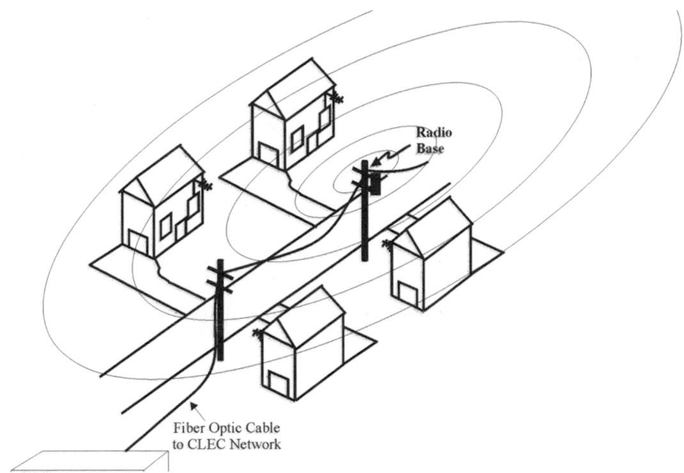

Figure 9.1 Wireless Local Loop System

wireless (cordless) telephones to use in the house and to use around the residential area where the WLL transmitters are located.

The most basic service offered by wireless local loop (WLL) systems is to provide standard dial tone service known as plain old telephone service (POTS). In addition to the basic services, WLL systems typically offer advanced features such as high speed data, residential area cordless service and in some cases, video services. To add value to WLL systems, WLL service providers will likely integrate and bundle standard phone service with other services such as cellular, paging, high speed Internet or cable service.

WLL systems can provide for single or multiple-line units that connect to one or more standard telephones. The telephone interface devices may include battery back up for use during power outages. Most wireless local loop (WLL) systems provide for both voice and data services. The available data rates for WLL systems vary from 9.6 kbps to over several hundred kbps. WLL systems can be provided on cellular and PCS, private mobile radio, unlicensed cordless, and proprietary wideband systems that operate in the 3.4 GHz range. The goal of international mobile telephone 2000 (IMT2000) standard is to provide 2Mbps service indoors. Unfortunately, there is a problem allocating bandwidth at the lower frequencies. With most of the frequencies assigned a limited bandwidth available at lower frequencies, there is limited room to give everyone the high bandwidth that is needed to provide 2 Mbps. This limits the ability to provide high speed data services such as digital video.

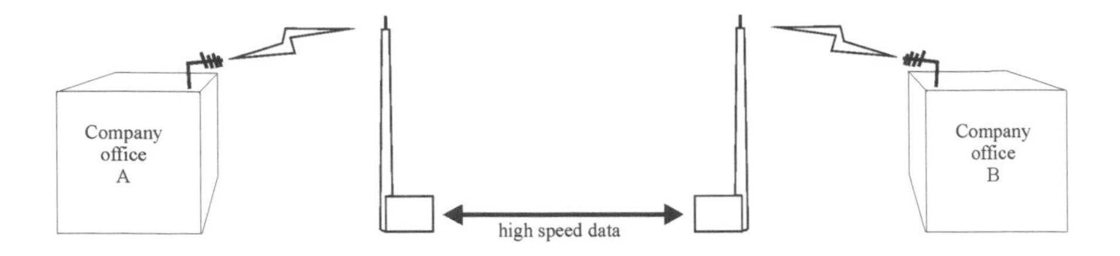

Figure 9.2 Wireless Bypass

High Speed Wireless Bypass

High speed wireless bypass service allows a company to use wireless service to bypass the need to use the telephone company's high speed data communication facilities. Wireless bypass has ordinarily involved the use of point-to-point microwave service to eliminate the need for leased lines. Wireless bypass service has changed to allow multipoint applications. Multipoint applications allow for advanced telephone services to several locations without the need to install a point-to-point microwave system.

Figure 9.2 shows a wireless bypass system that uses several microwave transmitters to provide service to several regions. These microwave transmitters are connected to each other by a high speed digital network. Each microwave antenna can serve one or more customers within a 5 to 10 mile radius with temporary or continuous service. Using this type of system, a company can have telephone or data services between offices without having to use the telephone company facilities. The data link between the microwave points may be owned by a company or may be leased from a telephone service provider.

Wireless Cable

"Wireless Cable" is a term given to land based (terrestrial) wireless

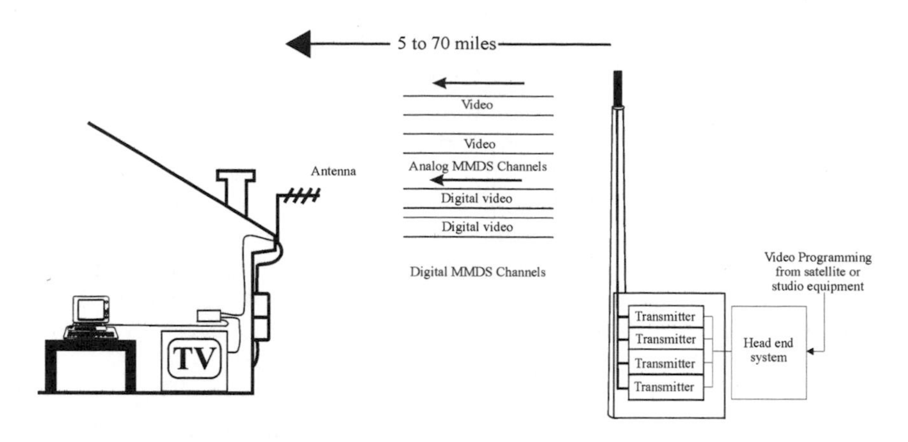

Figure 9.3 Wireless Cable System

distribution systems that utilize microwave frequencies to deliver video, data and/or voice signals to end-users. Wireless cable provides video programming from a central location directly to homes via a small antenna that is mounted on the side of the house. There are two basic types of wireless cable systems, Multichannel Multipoint Distribution Service (MMDS) and Local Multichannel Distribution Service (LMDS). In 1998, there were over 10 million MMDS wireless cable customers throughout the world and over 1.1 million in the United States.

In the 1970's, the first Multipoint Distribution Service (MDS) was the direct delivery of Home Box Office (HBO). Later, HBO moved to satellite transmission. The FCC reallocated many of these MDS frequency channels in the early 1980s and officially renamed them MMDS.

In 1996, some analog MMDS systems began upgrading to digital service. Through the use of digital video compression, digital transmission allows five to six times the video channel capacity. In addition to video programming, wireless cable can provide telephone service and data services.

Figure 9.3 shows that the major component of a wireless cable system is the head-end equipment. The head-end equipment is equivalent to a telephone central office. The head-end building has a satellite connection for cable channels and video players for video on demand. The head end is linked to base stations (BS) which transmits radio frequency signals for reception. An antenna and receiver in the home converts the microwave radio signals into the standard television channels for use in the home. As in traditional cable systems, a set-top box decodes the signal for input to the television. Low frequency MMDS wireless cable systems can reach up to approximately 70 miles. LMDS systems can only reach approximately 5 miles.

Wireless cable is one of the most economical technologies available for the delivery of pay television service. Wireless cable systems do not require extensive networks of cables and amplifiers, bringing the offered price generally lower than a traditional cable service. To the customer, a wireless cable system operates in the same manner as a traditional cable system. Because wireless signals are transmitted over the air rather than through underground or above-ground cable networks, wireless systems may be less susceptible to outages, offer better signal quality and be less expensive to operate and maintain than traditional cable systems. In conventional coaxial cable distribution networks, the television signal quality declines in strength as it travels along the cables and must be boosted by amplifiers thus introducing distortion into the television signal.

To add security for wireless cable systems, so unauthorized users do not gain access to the system (stealing service), signals from video sources are scrambled with a code. The user must have the code to successfully view the video signals. Like traditional cable systems, wireless cable systems employ "addressable" subscriber authorization technology, which enables the system operator to control centrally the programming available to each individual subscriber, such as a pay-per-view selection

There are two primary methods of providing a communication path back from the end customer to the network operator; a telephone line and wireless. Wireless cable systems have commonly only provided wireless downlink service (radio transmission from the system to the customer). Some of the new wireless cable systems now dedicate some of their radio channel capacity to uplink channels (from the customer to the system). This allows wireless cable systems to offer two-way service. Uplink channels typically are used to allow the customer to select programming sources (such as pay per view) or provide for two-way Internet access. Two-way service can also provide telephone service.

Wireless Local Area Network

Wireless Local Area Networks (WLANs) allow computers and workstations to communicate with each other using radio signals to transfer digital information. Wireless LAN systems may be completely independent or they may be connected to an existing wired LAN as an extension. While adaptable to both indoor and outdoor environments, wireless LANs are especially suited to indoor locations such as office buildings, manufacturing floors, hospitals and universities.

Wireless LANs provide all the functionality of wired LANs, but without the physical constraints of the wire itself. Wireless LAN configurations include independent networks, offering peer-to-peer connectivity, and infrastructure networks, supporting fully distributed data communications. Data rates for WLAN systems vary from 20 kbps to 60 Mbps.

Some wireless LANs also allow a personal area network (PAN). A PAN system normally covers the few feet surrounding a user's work space and provides the ability to synchronize computers, transfer files, and gain access to local peripherals.

Other types of wireless network systems include wireless metropolitan-area networks (WMANs) and wireless wide-area networks (WWANs). WMANs are private wireless packet radio networks often used for law-enforcement or utility applications. WWANs are wireless data transmission systems that cover a large geographic area using cellular or public packet radio systems. These wide area systems involve costly infrastructures, provide much lower data rates (regularly below 20 kbps) and often require users to pay for bandwidth on a time or usage basis. In contrast, on-premise wireless LANs require no usage fees and provide 100 to 1000 times the data transmission rate.

Figure 9.4 shows a typical wireless local area network (WLAN) system. In this diagram, several computers communicate data information with a wireless hub. The wireless hub receives, buffers and retransmits the information to other computers. Optionally, the hub may be connected to other networks (possibly other wireless network hubs) by wires.

Wireless Local Area Network (WLAN) radio systems use either narrow or wide radio channels or use infrared technology. A narrowband radio system requires coordination of different frequencies to avoid the possibility of interference from other similar frequencies. A wideband radio system uses a radio channel, which is much wider than is necessary to transfer the data information. The extra wide channel is used to spread the signal so it is less susceptible to interference. This is called spread spectrum technology. Infrared technology transfers

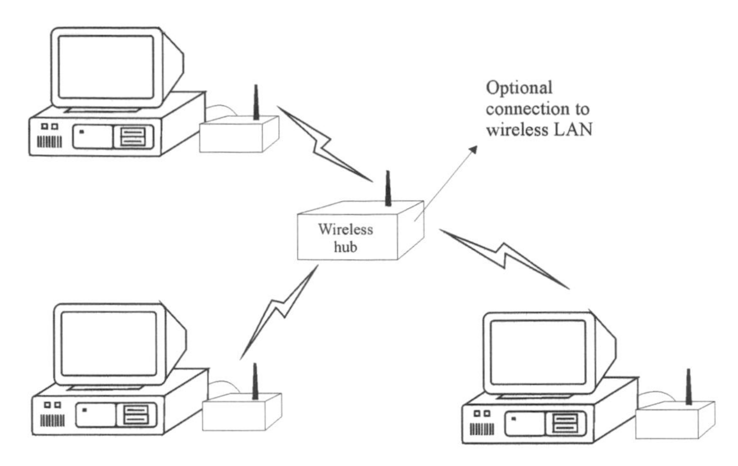

Figure 9.4 Wireless LAN System

data in the form of pulses of infrared light. Some short range infrared systems have shown data transfer rates of 60 Mbps.

The vast majority of wireless LAN products on the market use spread spectrum technology. Spread spectrum products operate primarily within the 900-megahertz and 2.4-gigahertz frequency bands, which do not require FCC licensing. They use limited transmitter power levels (less than one watt) and generally are designed to contain their signaling within 500-800 feet. Some WLAN systems operate in the 5.7 GHz spectrum. In February 1997, the United States FCC approved a plan to make available additional spectrum at 5.15-5.35 GHz and 5.725-5.875 GHz for use by a new category of unlicensed equipment. With these new global frequency bands, this will promote the development of many types of new devices (including WLAN) and improve the ability of manufacturers to compete globally by enabling them to develop products for the world marketplace.

Market Growth

The number of wired telephone lines worldwide in 1996 was 753 million and this is projected to grow to 1.2 billion by 2002 [1]. In 1998, the marketplace for fixed telecommunications services in the United States was estimated at $110 billion, with local telephony and data services accounting for approximately $47 billion [2]. Fixed wireless may be used to provide customers with a broad range of telecommunications services that compete with incumbent local exchange carriers ('ILECs'), cable television (CATV), other competitive local exchange carriers ('CLECs') and the interexchange carriers ('IXCs'). Predictions show that by the year 2005, wireless local loop is expected to soar to around 33 million lines in the United States.

The demand for wireless local loop is segmented into two key areas: services for developed countries and services for undeveloped countries. Developing markets are showing interest due to the demand for basic telephone service. Developing markets are expected to significantly outpace developed markets for WLL service. It is estimated that by the year 2005, developing markets will have 148 million WLL subscribers while developed markets have only 54 million WLL subscribers.

In developed countries, it takes value added services and a tie to mobility services to get interest in WLL. In the particular case of England, point-to-multipoint (PMP) wireless service made some inroads by providing high quality WLL for internet.

The demand for WLL in developing countries is projected to be concentrated in China, India, Indonesia, Brazil, and Russia. These five markets are projected to account for nearly three-quarters of demand by developing markets by the year 2005. North America will contribute 41 percent of developed market WLL subscribers, followed by Western Europe with 32 percent and Asia-Pacific, led by Japan, with 26 percent [3].

Another important factor in the growth of WLL is the potential conversion of existing residential and business telephone lines as well as new services that can be provided with wireless local loop technology. A good example of this is in rural areas where supplying wired service was previously not cost effective and wireless service can offer high speed Internet access. Figure 9.5 shows the projected worldwide market growth for fixed wireless service.

1996	1997	1998	1999	2000
2	12	40	52	60

In millions, worldwide

Figure 9.5 Fixed Wireless Market Growth

Technologies

The key technologies behind fixed wireless service include digital video compression, hybrid systems, interference avoidance and infrared communications.

Digital Video Compression

One of the key advances in fixed wireless is the use of digital video compression. The use of digital transmission allows the sharing of multiple service through a single high capacity digital radio channel. Although it is possible to deliver advanced services through analog video transmission, changes in interactive services and shared services can be implemented through software changes on a digital system.

Digital video usually requires a high data transmission rate. This would consume more bandwidth than analog video transmission, resulting in a lesser number of video channels. To increase the efficiency of digital video, video signals can be compressed using software programming. The video compression involves characterizing the video signal and sending the coded (compressed) digital video information by the radio channel. When it is received at the set top box, it is reconverted (decoded) into its original form.

There must be a standard conversion process to allow products that are manufactured by different companies to be capable of working together. The prevailing standard for digital compression was developed by the motion picture experts group (MPEG). There are various levels of MPEG compression; MPEG-1 and MPEG-2. MPEG-1 compresses by approximately 52 to 1. MPEG-2 compresses up to 200 to 1. MPEG-2 typically provides digital video quality that is similar to VHS tapes with a data rate of approximately 1 Mbps. MPEG-2 compression can be used for HDTV channels, however this requires higher data rates.

Hybrid Systems

Wireless services often cannot provide some or all of the service in both directions between the customer and the system. For example, fixed wireless television services do not commonly have the capability (nor do they need the capability) to provide a video channel back from home to the television station. However, some interactive response is required to activate or control some fixed wireless services. The use of hybrid systems such as the combination of wireless services from the system to the customer and telephone control from customer to the system allows many of the new services to be successful.

Number Portability

It is possible with recent telecommunication regulatory changes to allow the customer to keep their telephone number when their telephone service changes to a new local telephone service provider. This is called number portability.

Number portability has only been made recently possible because many of the wired telephone systems now use software to control the switching of calls to their destination. In the past, the destination was primarily determined by the location of the telephone. Now, calls are routed to switches by software control so this allows the routing of a dialed telephone call to be changed through the updating of software databases in the telephone switching control systems. Number portability supports competitive services for wire or wireless carriers and is important as a means to allow new telephone service providers to compete for new customers without disrupting customer businesses telephone numbering.

Interference Control

Radio interference is the unwanted effect from radio energy that produces a decrease in performance or other unwanted distortion. Many of the new wireless systems use independent (uncoordinated) wireless access with radio devices. The use of these systems can result in interference to or from nearby systems. Interference control procedures can be used to detect interference from other nearby devices and automatically change or control access to avoid the interference.

Interference control is an important area of consideration because of the potential to increase efficiency in spectrum through the use of idle radio spectrum by multiple devices. The increased efficiency allows the handling of higher bandwidths for fixed wireless devices. There are several ways to reduce interference including directional antennas, frequency hopping, or the use of spread spectrum radio technology.

Directional antennas are one method of interference control. Fixed wireless systems can use highly directional antennas to reduce or almost eliminate interference from nearby devices. Directional antennas can vary from simple physical designs (such as a satellite dish) to complex electrically controlled antennas.

Many of the new wireless LAN and some wireless local loop technologies operate in frequency bands that can be occupied by other users. Most of these systems use some form of spread spectrum technology to minimize the amount of interference. There are two commonly used spread spectrum techniques: frequency hopping and direct sequence.

With frequency hopping, an approach developed to prevent intrusion by outsiders, a radio transmission moves from frequency to frequency based on a pseudo-random algorithm. Because the time the transmission remains on any one particular frequency is just fractions of a second, it is very difficult for an intruder to jam or even recognize a transmission.

With direct sequence, the transmission, rather than hopping from frequency to frequency, is spread over the entire allowed band. Because both the transmitting and receiving equipment are programmed with the same spreading and despreading code, the signal can be reconstructed at its destination. Direct sequence techniques also reduce the effect of interference by other radio signals.

Infrared Communications

Infrared systems use very high frequencies, just below visible light in the electromagnetic spectrum, to carry data. Like light, IR cannot penetrate opaque objects; it is either directed (line-of-sight) or uses diffuse technology. Inexpensive directed systems provide very limited range (3 ft) and ordinarily are used for personal area networks (PANs), but are occasionally used in specific WLAN applications. High performance directed IR is impractical for mobile users and is therefore used only to implement fixed subnetworks. Diffuse (or reflective) IR WLAN systems do not require line-of-sight, but cells are limited to individual rooms.

InfraRed (IR), "optical" links can be configured in different ways, depending on orientation, and type of radiation antenna. A beam traveling directly (line of site) from the transmitter to the receiver, without reflection is called a directed signal. If there is no direct path, the signal can be reflected by the ceiling and walls. This is called diffuse reflection signaling. The directed, line-of-sight configuration is not capable of supporting one-to-many or many-to-one connections. For most practical implementations, it is not useful to consider a line of sight restriction. However, directed systems are useful in several applications, such as docking stations for laptop computers.

The most challenging configuration, but also the one which offers the most freedom, is the non-directed non-light-of-sight signaling. The transmitter sends signals in a wide angle to the ceiling and after one or several reflections, the signal arrives at the receiver. The maximum designed data rate for this system is 60Mbps. With diffuse reflection, the data rate depends on the room size. By using a complex mathematical model, it is possible to predict, for example, that the data rate in a 15x15 foot room will be 12Mbps.

Systems

Personal Access Communication System (PACS)

Personal Access Communication System (PACS) is a low mobility wireless telephone system which works closely with the landline telephone network. PACS was developed with influence from Wireless Access Communication System (WACS), a Bellcore [4] standard and on the Japanese Personal Handy Phone (PHP) which was defined in a standards document issued by the Japanese research center for radio (RCR) standards group.

The basic parts of a PACS system include cordless handsets and/or a home base station dialtone generator called a wireless access fixed unit (WAFU). The WAFU converts a PACS radio channel into a dial tone signal which allows the user to connect standard telephone devices, like answering machines or fax machines, to the PACS system. Nearby system base stations are called radio ports (RP). The RP is connected to a radio port control unit (RPCU) that is normally connected directly to the wired telephone network.

The PACS system uses a high quality 32 kb/s ADPCM speech coding processor to provide quality voice communications. This allows the PACS system to be used with moderate bit rate modems and fax machines. When the end customer requires a higher speed data connection, the PACS system can be adapted to provide 64 kb/s data transfer capability.

Basic Exchange Telephone Radio Service (BETRS)

BETRS is part of a United States FCC Report and Order that allows use of the spectrum for both mobile and WLL applications. Various technologies can be used for the service. Some of the BETRS systems in use are digital radio systems that operate at frequencies near 150, 450, and 850 MHz. These systems are used primarily in areas where wired telephone service is not economically viable. BETRS is sometimes referred to as Basic Exchange Radio (BEXR) Service.

Point-to-Multipoint (PMP) Microwave

Point-to-Multipoint (PMP) systems use hubs to connect several end users to high speed networks. PMP systems typically use microwave frequencies to provide high speed data services. The wide bandwidths available at microwave frequencies allow the use of data rates that approach fiber links. PMP wireless systems connect a high speed data user's directional antenna back into the wired world through a the hub. The hubs are connected to a backbone network that interconnects hubs to each other and to other networks (such as the public telephone network).

WinStar is one of several companies that provides high speed point-to-multipoint (PMP) wireless bypass telecommunications services to businesses and residential customers in urban areas in the United States. The WinStar system operates in the 38 GHz radio spectrum in more than 125 Metropolitan Statistical Areas ('MSAs') throughout the United States [5]. The WinStar system is a combination of high speed data wireless radios that are inter-connected to a fiber optic network.

The WinStar system is called "Wireless FiberSM." Each 100 MHz Wireless Fiber channel can support transmission capacity of one DS-3 at 45 Mbps. The WinStar's development of multipoint facilities that is planned to begin 1998 should allow a single 100 MHz Wireless Fiber channel to support one OC-3 equivalent of capacity at 155 Mbps.

WinStar identifies strategically located sites to serve as hubs in each of its metropolitan areas. These hub sites are connected via wireless fiber links to customer buildings. Certain characteristics of the 38 GHz frequency, including the effective range of its radio signal and the small amount of dispersion (i.e., scattering) of the radio beam as compared to the more dispersed radio beams produced at lower frequencies, allow for multiple hub sites using the same channel in a licensed area. Because the antennas are highly directional, it is possible to re-use several 38 GHz channels in a single geographic area.

Multichannel Multipoint Distribution Service (MMDS)

MMDS is a wireless cable service that is used to provide a series of channel groups, consisting of channels specifically allocated for wireless cable (the "commercial" channels). In the United States, MMDS service evolved from radio channels that were originally authorized for educational purposes. MMDS video broadcast systems have been in service since the early 1990's providing up to 33 channels of analog television over a frequency range from 2.1 to 2.69 GHz. Optionally, there are 31 "response channels" available near the upper end of the 2.5 to 2.69 band. These response channels were originally intended to transmit a voice channel from a classroom to a remote instructor.

MMDS systems normally transmit using an omni-directional antenna with the receivers, usually at a home, using a small directional antenna. A frequency down-converter is supplied by the service provider that converts the MMDS frequencies to ordinary VHF TV channels.

Although MMDS is a relatively established service, FCC rule changes in 1990 allowed single operators to license larger numbers of channels and operate them as a broadcast-like service, which made MMDS more economically viable. After this change in rules, a tidal wave of industry interest resulted in 24,000 license applications. Well over 160 MMDS systems were operating in the United States in 1997, with half of them serving rural areas. It is possible that "private" educational video transmissions will be widely distributed by optical fiber networks, rather than by MMDS services, keeping MMDS service offerings in check. MMDS is currently used to enrich junior college classroom offerings and used for adult evening classes at remote locations.

Local Multipoint Distribution Service (LMDS)

Local multipoint distribution service (LMDS) provides similar service as MMDS at higher frequencies. Because of the use of high frequency that is attenuated quickly, LMDS must install antennas approximately every five miles. When antennas are installed, this offers tremendous system capacity possibilities. Digital LMDS has the capacity for thousands of video, audio and data channels. MMDS and LMDS both will enable wireless cable systems to transmit high definition television (HDTV) over their existing allocated frequencies and channels. LMDS has more than twice the bandwidth of AM/FM radio, VHF/UHF television, and cellular telephone combined. LMDS may prove to be the shortest path to mass delivery of two-way video and high-speed data services ever commercially offered.

Several LMDS systems are in operation. These systems are analog and several digital systems are in development. CellularVision, a company that provides service to New York City has been providing service since 1996. Recently, Cellular Vision offered wireless Internet service to consumers that allowed data rates of 500 kbps. CellularVision also announced that it plans to provide wireless Internet to businesses with data rates up to 32 Mbps. Other companies that are developing LMDS technology include Hewlett Packard, Texas Instruments, Stanford Telecom and others.

Initially, radio coverage was perceived to be a challenging factor for LMDS service as microwave signals cannot regularly penetrate into buildings very well. This would have limited the deployment of LMDS systems into large cities such as New York City. However, because microwave signals reflect off large objects (such as buildings), CellularVision claims it has been exploiting the reflections off buildings to provide extended coverage.

The key challenge for LMDS operators will likely be the initial construction of systems. Unlike cellular telephone operators, which were able to start with large cells and gradually split them into smaller cells, LMDS systems can only offer small radio coverage areas (cells). This means that LMDS service providers will likely target areas with a large concentration of potential users.

LMDS frequencies in the United States are near 28 and 29 GHz downlink and near 31 GHz for an uplink return path (transmission from the end customers equipment to the system operator).

IEEE Wireless LAN

There are many wireless LAN (WLAN) systems in use worldwide. Many of these systems use radio communications protocols that are proprietary to the manufacturer of WLAN equipment. This has resulted in incompatibility between products. To overcome this limitation, the Institute of Electrical and Electronic Engineers (IEEE) created an industry standard specification in June 1997 (802.11) that defines a wireless protocol for homogeneous systems. The IEEE 802.11 standard for wireless LANs (WLANs) allows interoperability among products from different vendors.

The 802.11 specification has three transmission options; one infrared (IR) option and two radio frequency (RF) options: direct sequence spread spectrum (DSSS) and frequency hopping spread spectrum (FHSS).

When choosing a wireless LAN system, customers should consider their future requirements and how the system can be expanded or changed to meet the needs of their future requirements. An Infrared (IR) system provides a basic 1 Mbps data transfer rate with an option for a 2 Mbps system. The Radio Frequency systems can provide both 1 Mbps and 2 Mbps data transfer rate. Work is underway to standardize systems that have higher data transfer rates.

Services

Fixed wireless systems are capable of offering a full range of telecommunications services, including local telephony, long distance, high-speed switched data and dedicated services. Local telephony services offered by fixed wireless primarily use digital switches to route voice transmission anywhere on the public switched telephone network. Fixed wireless data transmission services include Internet access and WAN services using frame relay, IP and ATM data transport protocols.

Local Telephony Services

Fixed wireless service can provide local dial tone and connection to both regional and long distance calling. Basic and enhanced voice services include analog and digital (ISDN) voice-only telephone lines to customers. Other features are call waiting, call forwarding, conference calling and voice mail, as well as operator and directory assistance services. Fixed wireless systems usually provide full switched toll grade quality voice service. The customer may use standard telephone devices that connect to RJ-11 by using a converter so they will not need to change any of their existing telephone devices.

The thinking of the Ad Hoc WLL Group of Telecommunications Industry Association (TIA) was to use 64kbps (a single DS0 voice channel) as a basic wireless local loop data rate. Users can then combine n x 64 kbps as needed. The use of a 64 kbps basic channel data rate allows WLL to be compatible with ISDN designs. The provision of 144 kbps basic rate ISDN lines (two 64 kbps and one 16 kbps signaling channel) allows fixed wireless customers to perform multiple functions such as simultaneous voice and data links for moderate speed Internet access.

Long Distance Telephone Services

Wireless service providers can offer long distance services to its customers. The wireless service provider can resell long distance through agreements with inter-exchange carriers (IXCs). If the wireless service provider can bypass the point of presence (POP) with the local exchange company (LEC), it is possible that they may not be required to pay the tariffs (up to 45% of revenue) to the LEC for access to their customers. This provides a significant cost advantage to the wireless service provider.

Centrex Services

Centrex is a service offered by local telephone service providers that allows the customer to have features that are typically associated with a private branch exchange (PBX). These features include three or four digit dialing, intercom features, distinctive line ringing for inside and outside lines, voice mail waiting indication and others. Centrex services are provided by the central office switching facilities in the telephone network.

Many of the fixed wireless systems allow the wireless service provider to offer Centrex type services. Business customers can use fixed wireless as their primary Centrex provider, as a supplement to the LEC's Centrex service, or as an addition to customer-owned PBX. Centrex services provide businesses with access to the local, regional and long distance telephone public networks.

Data Services

Fixed wireless service providers can provide high speed data services with data rates that range from 9.6 kbps up to 155 Mbps. Low speed data services are commonly offered by wide area public networks such as cellular or packet data networks and high speed data rates are available through fixed point systems such as LMDS.

The growth of the LAN, WAN, and Internet industries is causing data flow to become an increasing portion of overall telecommunications traffic. The ability to quickly access and distribute such data and other information is critical to business, education and government entities.

If the wireless service provider covers a relatively large geographic area, it is possible to offer wide area network data transport services. This can be in the form of low speed packet data networks such as X.25 or high speed virtual LAN networks, frame relay, switched multi-megabit data service (SMDS) or other types of data networks. Some of the fixed wireless technologies permit broadband (high speed) data services. These services allow interactive video applications, such as video conferencing, appropriate for highly customized commercial and institutional demands.

Internet Access

By using the data service capability, fixed wireless can offer dedicated or dial-up Internet access services, as well as web hosting and Intranet services. The wireless operator can partner with an Internet service provider (ISP) or become their own ISP.

Video

Fixed wireless networks may be able to provide the user with analog or digitized video services along with the provisioning of interactive video services. Compressed digital video allows many of the fixed wireless services to offer 50 or more video channels. Some fixed wireless systems can deliver video on demand (VOD) services.

Typical service fees for video services range from approximately $20 per month for basic television programming to $50 per month for a full package of enhanced video channels such as HBO and Showtime. If the wireless service provider becomes a distributor of video services, there will be a cost for the programming content from the providers of the video programming.

Future Enhancements

There is a lot of confusion about what fixed wireless is supposed to offer. Some fixed services are tied to mobility and others offer variable data rates with their maximum data rate approaching broadband data speeds.

Multiple Services

There are several wireless systems in development that combine fixed wireless services with mobile wireless. A good example of a multiple service system is the Universal Mobile Telecommunications System (UMTS) that is in development in Europe. The UMTS system objectives include the offering of personal telecommunications services that use the combination of wireless and fixed systems to provide seamless telecommunications services to its users. This systems will combine paging, mobile voice (cellular), satellite, high speed data and other wireless systems to provide a standard solution for wireless devices. UMTS will allow on-demand transmission capacities of up to 2 Mb/s in some of its radio locations. It should be also be compatible with broadband ISDN services.

1. In 1997, Bellcore was the research and development organization of

the 7 US regional Bell operating companies.
[2]. Globalstar 10K report, 31 March 1998, US Securities and Exchange Commission.
[3]. 10K report, 31 March 1998, US Securities and Exchange Commission
[4]. "Wireless Resale Market Report", Multimedia Publishing Corporation, Houston, TX, 1997, pg. 301.

Chapter 10

Television and Radio Broadcast

Introduction

Radio broadcast services have been able, over the past 100 years, to bring news and entertainment to listeners without wires. Radio broadcast systems use wireless technology to transmit the same information to many recipients. The most popular applications of broadcast services are television and audio (commonly called radio) transmission. Radio and television broadcast services, as we know them, are now changing from analog to digital service to allow for new services and more individualized features.

For many years, broadcasters had monopolized the distribution of some forms of information to the general public. This had resulted in strict regulations on the ownership, operation and types of services radio broadcast companies could offer. Due to the recent competition of wide area information distribution, Congress eased its regulation of the broadcast industry through the Telecommunications Act of 1996.

In the mid 1990's, a major shift occurred in the broadcast industry; the conversion from analog systems to digital systems. The use of digital transmission allows a broadcaster to bundle multiple types of services onto a single radio channel signal. This allows radio broadcast companies to offer various types of services including digital television, paging and other information services. The ability to integrate several services into one transmission signal typically reduces the average equipment cost per customer.

Because most broadcast systems can be listened to by anyone with a suitable radio receiver, revenues from broadcast companies primarily come from advertisers. Over the past 10 years, advanced control systems have been developed that allow control of broadcast signal reception and decoding. These systems allow broadcasters to receive revenues from customers on a service subscription basis (e.g. pay per view).

Television

The technology that is used for television broadcast was developed over 50 years ago. The success of the television marketplace is due to standardized, reliable, and relatively inexpensive television receivers and a large selection of media sources. The first television transmission standards used analog radio transmission to provide black and white video service. These initial television technologies have evolved to allow for both black and white and color television signals, along with advanced services such as stereo audio and closed caption text. This was a very important evolution as new television services (such as color television) can be on the same radio channel as black and white television services.

While television technology is proficient at distributing analog video and audio signals, it does not readily allow the sending and receiving of digital data. Several new television broadcast technologies are currently proposed to deliver high quality video and audio as well as information services using digital signal transmission.

Using the existing television and radio broadcast frequencies, new technologies allow transmission of high-definition television (HDTV). HDTV is the term used to describe a high resolution video and high quality audio signal. These systems can be analog or digital. Digital HDTV systems can also provide data services. The current system used in the United States offers several channels of high resolution video and compact disk (CD) quality surround sound audio.

In 1996, the changes made to the United States Telecommunication Act altered many rules for the regulation of television companies. In the past, television companies were restricted on how many transmitter stations they could own and how many people these radio stations could reach. The major effect of the changes was to allow television broadcast companies to own more transmitters and serve more customers. Since the passing of the Telecommunications Act of 1996, there have been many purchases of television stations in the US. These include the purchase of New World Communications by Rupert Murdoch's News Corporation (which owns and operates the Fox Television Network). This in turn created a group of 22 stations reaching 35% of the U.S. population. The Tribune's purchase of Renaissance Communications resulted in the creation of 16 television stations reaching 25% of the U.S. population. Sinclair Broadcasting's purchase of River City Broadcasting created a group of 29 television stations reaching 9% of the U.S. population.

A key issue for television broadcasters is how the conversion of analog television channels will change to HDTV or digital channels. Some of the new HDTV television technologies allow the co-existence of analog and digital television on the same radio channel. This would be similar to the addition of color signals to the original black and white television signals that allowed existing black and white televisions to continue to receive black and white television while new color televisions could view color broadcasts. Unfortunately, the combination of analog and digital signals on the same radio channel is complex and this process does not offer the most efficient use of the radio channel bandwidth. The optimal digital television channel would replace the NTSC standard television signals with a digital signal. If this occurs, existing television sets could not receive the digital signal without a converter box. The converter box would change the digital signal to a standard analog signal.

Some additional radio spectrum for digital television and high-definition television may be awarded. However, television channels require a large amount of bandwidth and much of the radio channels (radio spectrum) have already been assigned to other users. In the United States, it is anticipated that some of the existing television broadcast spectrum may be auctioned off in 2002 with the requirement of analog broadcasting to cease operation in 2005.

Figure 10.1 shows a typical television system. A television system consists of a television production studio, a high power transmitter, a communications link between the studio and the transmitter, and net-

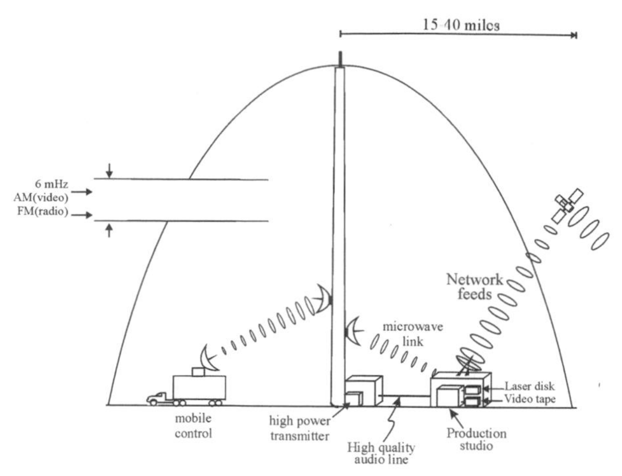

Figure 10.1 Television System

work feeds for programming. The production studio controls and mixes the sources of information including video tapes, video studio, computer created images (such as captions) and other video sources. A high power transmitter broadcasts a single television channel. The television studio is connected to the transmitter by a high bandwidth communications link that can pass video and control signals. This communications link may be a wired (coax) line or a microwave link. Many television stations receive their video source from a television network. This allows a single video source to be relayed to many television transmitters.

Audio Broadcasting ("Radio")

Communities rely on audio ("radio") broadcast information for weather advisories, economic information and public safety and service announcements. Radio broadcast systems have been used for over 100 years. The two dominant radio broadcast systems use analog transmission in the form of amplitude modulation (AM) or frequency modulation (FM). There are new proposed radio technologies that promise to add digital audio and data services to conventional radio broadcasting.

Radio broadcasting companies in the U.S. radio industry are also consolidating as a result of the 1996 Telecommunications Act. The FCC's new radio ownership rules allow group owners to own or operate a maximum of eight stations in a market with more than 45 stations, seven stations in a market with between 30 and 44 stations, six stations in a market with 15 to 29 stations, and five stations in markets with less than 15 stations. As of July 1, 1996, the largest radio group (in terms of market reach and revenue) resulted from the combination of the radio stations owned by CBS, Group W, and Infinity Broadcasting under Westinghouse Electric Corp. The new group includes 82 stations reaching an average of 2.6 million listeners per quarter hour.

The efficient use of FM radio channels has resulted in some of the radio channel bandwidth being unused. With modifications to the transmitter, it is possible for FM broadcast stations to simultaneously send some additional information (sub-channels) with their audio broadcasts. Sub-channels can contain audio or digital information. Chapter 7 discussed the use of sub-channels for paging services. Additional services include low bandwidth audio broadcasts and private messaging services.

Figure 10.2 shows a typical radio broadcast system. The radio broadcast system consists of a production studio, a high power AM or FM transmitter, a communications link between the studio and the transmitter, and network feeds for programming. Similar to television broadcast systems, radio broadcasting involves the use of various types of information sources called "program sources." These program sources come from compact discs, tape recordings, soundproof audio studios, remote location sites (such as a van), or other network sources. The production studio controls and mixes the sources of information including audio compact discs, audio studio, audio tape and other audio sources. A high power transmitter broadcasts a single radio channel. The studio is connected to the transmitter by a coaxial

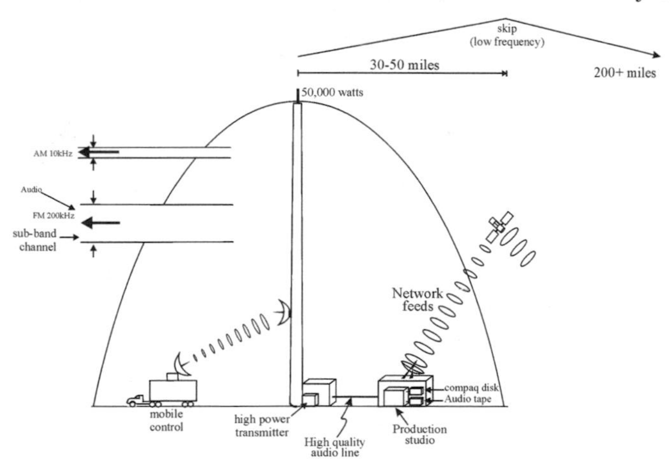

Figure 10.2 Radio System

cable, special leased telephone line (extra high quality), or dedicated radio link. Many radio broadcast stations receive their programming source from a radio broadcast network. This allows a single audio source to be relayed to many radio broadcast transmitters. The diagram also shows how a sub-channel is combined to provide a private audio broadcast service.

Market Growth

Television has been embraced by almost everyone in the United States. The first television sets in the United States were put into operation in 1946. By 1949, receiver sales were exceeding 10,000 per month. In 1995, there were over 160 million television sets operating in the United States.

It would be difficult to find anyone who does not already own a TV. In fact the latest statistics report that over 98% of all American households have at least one TV, with 70.9% of American households owning two or more television sets, and 25% have three or more sets! By 1996, color television sales were over 20 million units. Of those sales, nearly 90% are replacement or additional purchases. Figure 10.3 shows the growth of the color television market in the United States over the past 10 years.

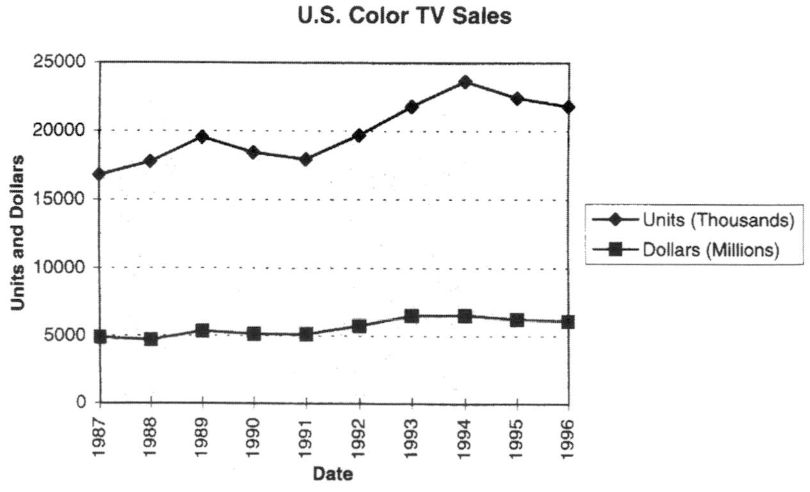

Figure 10.3 Color Television Market Growth
Source: EIA Electronic Market Data Book

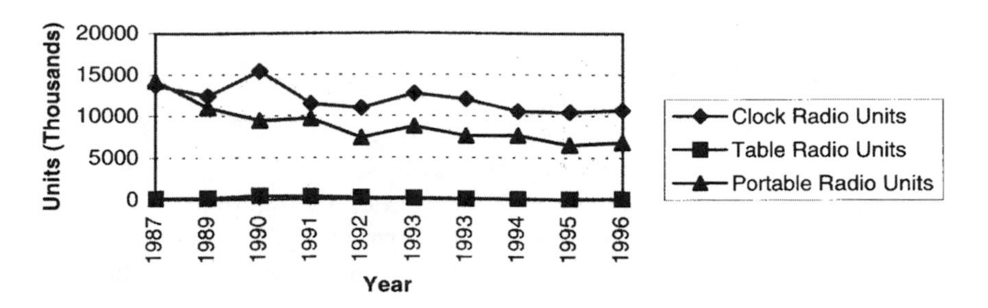

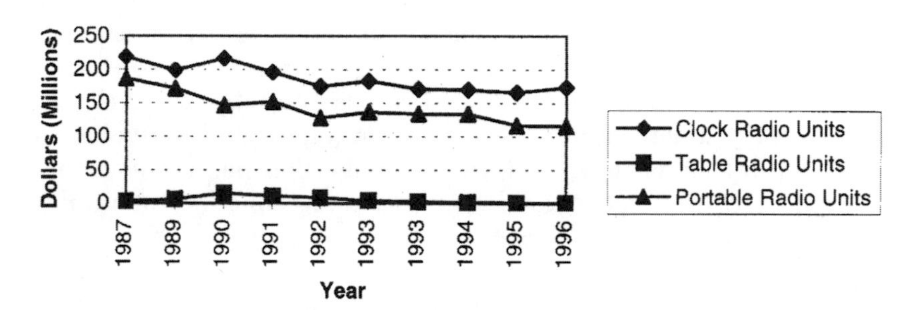

Figure 10.4 Radio Market Growth
Source: EIA

Despite the growth of other media, including television, radio has continued to flourish since its launch 100 years ago. Today there are approximately 12,000 radio stations in the country. Of these 85% are commercial stations and the balance are non-commercial or public stations.

Over 210 million people listen to the radio each day. Of these 60% listen to FM only, 8% listen to AM only, with 28% listening to both AM and FM. The most popular radio music format is Country, followed by News-Talk radio. Figure 10.4 shows the growth of the radio market of the past 5 years. This figure does not include car radios.

Technologies

Analog Video

Analog video contains a rapidly changing signal (analog) that represents the luminance and color information of a video picture. Sending a video picture involves the creation and transfer of a sequence of individual still pictures called frames. Each frame is divided into horizontal and vertical lines. To create a single frame picture on a television set, the frame is drawn line by line. The process of drawing these lines on the screen is called scanning. The frames are drawn to the screen in two separate scans. The first scan draws half of the picture and the second scan draws between the lines of the first scan. This scanning method is called interlacing. Each line is divided into pixels that are the smallest possible parts of the picture. The number of pixels that can be displayed determines the resolution (quality) of the video signal. The video signal breaks down the television picture into three parts: the picture brightness (luminance), the color (chrominance), and the audio.

There are three primary systems used for analog television broadcasting: NTSC, PAL and SECAM. The National Television System Committee (NTSC) is used for the Americas, while PAL and SECAM are primarily used in the UK and other countries. The major difference between the analog television systems is the number of lines of resolution and the methods used for color transmission.

There have been enhancements made to analog video systems over the past 50 years. These include color video, stereo audio, separate audio programming channels, slow data rate digital transfer (for closed captioning), and ghost canceling. The next major change to television technology will be its conversion to HDTV.

Figure 10.5 demonstrates the operation of the basic NTSC analog television system. The video source is broken into 30 frames per second and converted into multiple lines per frame. Each video line is sent by starting the radio signal with a burst pulse (called a sync) followed by a signal which represents color and intensity. The time relative to the starting sync is the position on the line from left to right. Each line is sent until a frame is complete and the next frame can begin. The television receiver decodes the video signal to position and control the intensity of an electronic beam that scans the phosphorus tube ("picture tube") to recreate the display.

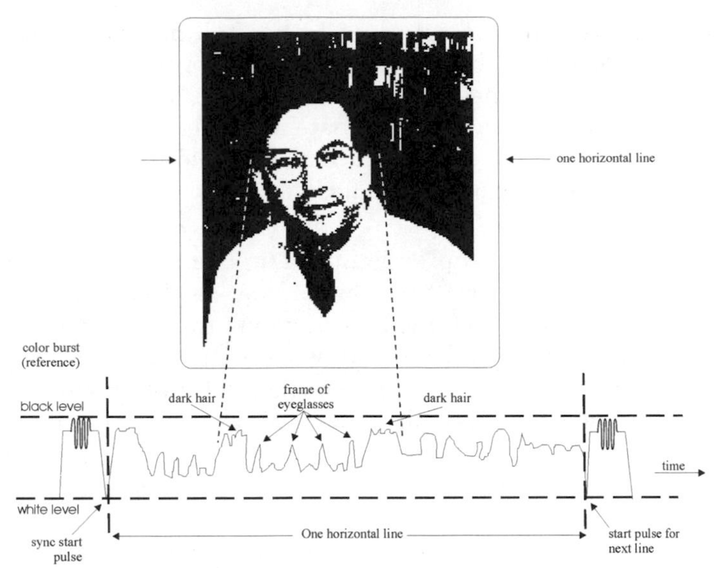

Figure 10.5 NTSC Analog Video

Digital Video

Digital video is the sending of a sequence of picture signals (frames) that are represented by binary data (bits) that describe a finite set of color and luminance levels. Sending a digital video picture involves the conversion of a scanned image to digital information that is transferred to a digital video receiver. The digital information contains characteristics of the video signal and the position of the image (bit location) that will be displayed. Digital television continues to send information in the form of frames and pixels. The major difference is the frames and pixels are represented by digital information instead of a continuously varying analog signal.

The first digital television broadcast license for the United States was issued to a Hawaiian television station in September 1997. Digital television sends the video signal in digital form. Ironically, many television signals have been captured and stored in digital form for over 10 years. These digital video signals must be converted to analog video signals for transmission to standard NTSC analog televisions.

Most digital video signals commonly use some form of data compression. Data compression involves the characterization of a single picture into its components. For example, if the picture was a view of the blue sky, this could be characterized by a small number of data bits that indicate the color (blue) and the starting corner and ending corner. This may require under 10 bytes of information. When this digital information is received, it will create a blue box that may contain over 7,200 pixels. With a color picture, this would have required several thousand bytes of information for only one picture.

In addition to the data compression used on one picture (one frame), digital compression allows the comparison between frames. This allows the repeating of sections of a previous frame. For example, a single frame may be a picture of city with many buildings. This is a very complex picture and data compression will not be able to be as efficient as the blue sky example above. However, the next frame will be another picture of the city with only a few changes. The data compression can send only the data that has changed between frames.

There are several digital video compression systems. The most common form of digital video compression of video signals conforms to the motion picture experts group (MPEG). There are various levels of MPEG compression; MPEG-1 and MPEG-2. MPEG-1 compresses by approximately 52 to 1. MPEG-2 compresses up to 200 to 1. MPEG-2 ordinarily provides digital video quality that is similar to VHS tapes with a data rate of approximately 1 Mbps. MPEG-2 compression can be used for HDTV channels, however this requires higher data rates.

Digital television broadcasting allows for "multicasting" (simultaneously sending) several "standard definition" television channels (normally up to five channels). Unfortunately, high definition digital television channels require a much higher data transmission rate and it is likely that only a single HDTV channel can be sent on a digital television channel.

AM Radio

AM radio is the broadcasting of audio signals through the use of an amplitude modulated radio carrier signal. Most AM radio broadcast systems use relatively low frequency radio signals (compared to FM radio and television). Amplitude modulation offers relatively efficient use of the radio bandwidth. Unfortunately, AM modulation is sensitive to radio frequency distortions caused by electronic noise such as lightning, car ignition sparks, power lines and other forms of amplitude distortion.

Because many AM radio stations transmit at lower frequencies, this sometimes results in the long distance skipping of the radio wave, particularly at night. Skipping occurs when the radio signal is bounced back from the sky (called refraction) to the ground. AM skipping can occur over hundreds of miles. This limits the use of the same AM radio channels to cities that are separated by hundreds of miles.

Several years ago, the AM radio broadcast system was updated to include AM stereo transmission. Over 1,000 AM stations use quadrature amplitude modulation (QAM) to transmit stereo signals. This AM stereo transmission system does not interfere with the standard reception of mono (single channel) AM signals. Other AM transmission systems have been demonstrated that virtually eliminate the interference noise that is common to AM signals.

FM Radio

FM radio is the broadcasting of audio signals (and in some cases sub band audio and data signals) through the use of a frequency modulated radio carrier signal. Frequency modulation offered a major improvement in radio broadcast audio quality. Since its commercial introduction, FM radio broadcast transmission has evolved to include FM stereo and FM subcarrier data transmission.

FM radio modulation is not as susceptible to the same types of distortions as AM radio. This is because the amplitude spikes that occur from lightning bursts, ignition noise and power lines do not change frequency in a way that is regularly detected by a FM receiver. The result is a robust quality audio signal.

FM broadcast radio transmission involves the sending of both the mono (combined single channel) audio and stereo (2 channel, left and right audio). The first modulated part of the FM signal contains the information from both the left (L) and right (R) audio channels (L + R). To create stereo, another audio channel is added in the same channel at 38 kHz, above and below the radio carrier signal center frequency, that contains the difference between the left and right channels (L - R). If an FM receiver has the capability of stereo, it will process the sum and difference signals to extract the left and right channels. If it does not have stereo capability, it will only use the initial sum signal.

Typical FM broadcast signals do not use their full authorized radio channel bandwidth when broadcasting. This results in a small portion of the authorized bandwidth being unused. The additional unused bandwidth may be used to broadcast additional services such as radio broadcast data system (RBDS). Such FM subcarrier systems send a very low bit rate data stream for information content, paging services and limited bandwidth audio broadcasts and other services. These data services can transfer about 1200 bits per second while new FM high-speed data services (FM HSS), such as DARC - SWIFT, can provide for data services between 8 to 16 kbps.

Recent innovations to radio broadcast technology include new FM high-speed subcarrier (FM HSS) services. There have been several prototype FM HSS systems tested by the national radio systems committee (NRSC) for industry standardization. The results of these tests will determine which systems are to be recommended for use in the United States. The ability of radio broadcasters to transmit high speed digital signals will allow many new types of types of services. Available in laboratories, but not yet ready for NRSC testing, are high speed subcarrier systems providing up to 56 kbps to fixed receivers.

Digital Radio

Digital radio broadcast service provides digital audio and other services on the radio broadcast channels. The conversion of radio broadcast systems to digital for radio broadcast can be implemented in several ways. One way is the combination of analog and digital signals on the same radio channel. The alternative method other way is to totally replace the analog signal with a digital signal.

In the United States, the combined solution is called in-band, on-channel (IBOC) and it is the approach preferred by the broadcasters. There are two ways to convert a radio broadcast signal to digital transmission for IBOC. The first method adds the digital data to the existing analog signal before it modulates the radio signal and the second method adds a separate radio signal in an unused portion of the FM broadcast channel. Both of these system types allow the existing analog broadcast service to continue and are completely compatible with conventional AM or FM broadcasts. So far, however, they only provide additional data or lower fidelity audio services. These new digital services are available only to customers that have a receiver that is capable of decoding and processing the digital data.

Completely digital radio, also known as digital audio broadcast (DAB), has been implemented on satellite systems at much higher frequencies than the standard radio broadcasts. In Europe and Canada, terrestrial DAB implementations at L-band frequencies have begun to be installed for testing and demonstration purposes. These systems will require listeners to have completely new radio receivers, however, the same receivers may be able to be used for satellite and terrestrial DAB reception.

Systems

National Television Standards Committee (NTSC)

The NTSC system is an analog video system that was developed in the United States and is used in many parts of the world. The NTSC system uses analog modulation where a sync burst precedes the video information. The NTSC system uses 525 lines of resolution (42 are blanking lines) and has a pixel resolution of approximately 148k to 150k pixels.

The NTSC system uses 6 MHz wide radio channels that range from 54 MHz to 88 MHz (for VHF channels 1-6), 174 MHz to 216 MHz (for VHF channels 7-13) and 470 MHz to 806 MHz (for UHF channels 14-69). Initially, the frequency range of 806 MHz to 890 MHz was available for UHF channels 70 to 83. The FCC reallocated these channels for cellular and SMR use in 1983.

When used in the United States, NTSC systems have a maximum transmitter power level that varies from 100 kWatts for low VHF channels (1-6), 316 kWatts for high VHF channels (7-13) to 5 million Watts for UHF channels (14-69). Television transmission limits are also established based on the class of service (local or wide area) for the authorized television broadcast company.

Phase Alternating Line (PAL)

The PAL system was developed in the 1980's to provide a common television standard in Europe. PAL is now used in many other parts of the world. The PAL system uses 7 or 8 MHz wide radio channels. The PAL system has 625 lines of resolution. Although PAL and NTSC systems are similar in function, they are not compatible. A converter box is required between the two systems.

High Definition Television (HDTV)

HDTV television offers more resolution than NTSC or PAL television systems. HDTV systems can be either analog or digital systems. High definition television has been offered in several countries since its introduction in Japan in 1988. The first US HDTV receivers were introduced at the 1998 Winter Consumer Electronics Show in Las Vegas.

HDTV radio broadcast channels can use the same 6 MHz channel bandwidth. However, it must replace the existing NTSC signal with a new high resolution analog or high speed digital radio signal. Initial demonstrations of HDTV required 2 standard television channels. The FCC has finally approved the "Grand Alliance" standard for high-definition television for the United States that only requires one standard television channel to send a HDTV digital channel and supplementary services.

The FCC introduced, in April of 1997, a new table of digital television channel numbers and RF power level assignments for existing full-power television stations in the U.S. The new assignments were designed to give each television station coverage comparable to the station's existing radio coverage area when they convert to digital transmission.

The change in channel numbers is likely to be a significant challenge for television stations and many stations, especially in Southern California, have a reduced coverage area. This has resulted in the contesting of the new assignments by some television broadcasters. The Association of Maximum Service Telecasters (an association of local television stations) has proposed an alternative table of channel assignments to address the issues of established broadcasters.

The digital technology that allows high-definition television broadcasts in the U.S. can also be used for "multicasting," that is, transmitting up to five channels of "standard-definition" television programming. Many broadcasters are examining multicasting as an alternative to high-definition television. If the ability to provide more video channels is more desirable than providing high-definition broadcast quality video, HDTV broadcast service and products may be delayed for their entry in the US marketplace.

In July 1996, WRAL in Raleigh, North Carolina became the first U.S. television station to commence broadcast of high-definition television signals. As of early 1998, more than a dozen stations have licenses for digital transmission, and the number of licenses is increasing every month. HDTV is likely only to be available in the largest markets in the U.S. for at least the first year the service is provided, so viewers in smaller markets may have to wait many years before they have the opportunity to use digital and/or high-definition television receivers.

The data transmission rate of the HDTV system is 19 Mbps and it uses the motion pictures experts group (MPEG-2) video compression format. To allow for a gradual migration to HDTV service, HDTV transmission will also contain regular programming of standard television on HDTV radio channels. The simulcast transmission will continue for up to 15 years as standard NTSC televisions and transmitting facilities are phased out. Initially, HDTV receivers will have the capability to receive and display regular NTSC broadcasts.

AM Broadcast

AM broadcast in the US uses 10 kHz wide channels in a frequency range of 535 kHz to 1,605 kHz (107 channels). Transmission power levels for AM broadcasters is divided into four classes (class A - D). Class A allows the use of high power transmitters (10-50 kWatts of power) for large cities (hundreds of miles of coverage), class B is for primary cities or large rural areas (.25-50 kWatts), class C is used for local or rural service to communities (1 kW or less) and class D can use the A or B maximum power levels but cannot transmit at night due to skipping problems.

FM Broadcast

FM broadcast in the US uses 200 kHz wide channels in a frequency range of 88 MHz to 108 MHz (100 channels). Like AM transmitters, the transmission power levels for FM broadcasters are divided into 3 classes (class A - C). Class A is used to service small communities (up to 6 kWatts), class B allows the use of medium power transmitters (up to 50kWatts of power) for approximately 50 miles of coverage. Class C service is used for large cities (up to 100 kWatts) to provide service to almost 100 miles.

Digital Audio Broadcast (DAB)

A new type of digital radio broadcast technology has been proposed for use in the United States that can provide digital CD quality audio on a FM radio channel. For the in-band, on-channel (IBOC) approach proposed for the United States, digital audio broadcast (DAB) systems combine a digital signal with conventional FM audio signals. This ability for DAB to coexist with the conventional signal is important to allow the introduction of new digital services while continuing to provide broadcast services to existing customers that only have FM receiver capability. In Europe and Canada, DAB is being implemented at a much higher frequency band so that existing radio broadcasts will not be affected at all by DAB.

Services

Television

Television broadcast usually involves the sending of video information to a large group of consumers. Television service has been changing from free broadcast services (paid for by the advertisers) to multicast information services (some services paid for by the subscriber). A majority of revenue for television broadcast comes from advertising.

According to A.C. Nielson, the average daily viewing per TV household is 7 hours and 15 minutes, with men watching just over 4 hours, women watching about 4.5 hours, teens watching 3 hours and children watching 3.5 hours daily.

In millions	1995	1996	1997	1998
Network	$11,600	$13,000	$14,300	$15,215
National Spot	$9,100	$9,800	$10,380	$11,030
Local Spot	$9,980	$10,950	$11,715	$12,475
Syndication	$2,000	$2,215	$2,480	$2,675
Cable Network	$3,500	$4,470	$5,350	$6,060
Cable Local	$1,570	$1,960	$2,450	$2,775
TOTAL	$37,750	$42,395	$46,675	$50,230

Figure 10.6 Television Advertising Revenues
Source: Television Bureau of Advertising

Television revenues represent about 23% of all media generated revenue, including radio, magazines, newspapers and telephone directories (yellow pages).

Revenues for television advertising are growing approximately 5-15% annually with revenues expected to reach $50 billion by 1998. Cable advertising revenues show the largest annual gains, going from just over $230 million in 1982 to almost $9 billion in 1998. The growth in estimated annual US advertising revenues is as follows:

Rates vary widely for television advertising, depending on which type of advertising is chosen. These types include network, (where air times are chosen during specific shows, at a certain time, per commercial), spot (where run times are inserted during unsold network time slots), local (where commercials run during unsold network time slots) and cable. Pricing can range from $30 million for a 30 second spot during the SuperBowl, to $30 per 60 second spot during a local cable show.

The new digital television system allows the sending of several digital video sources (channels) on a single television broadcast channel. Each of these digital channels can be coded so only authorized customers can view them. This will allow broadcasters to provide for many new broadcast and multicast information services.

Audio

Radio ("audio") broadcast service is the sending of information (commonly audio) to a large group of consumers. AM and FM broadcast service has evolved to include subband services such as paging and private audio channels. Like television, a majority of revenue for audio broadcast comes from advertising.

Radio broadcast revenues come from advertising, which has tripled over the last 10 years, reaching over $15 billion per year. Most commercial stations receive the bulk of their ad revenues from local advertising, as opposed to television, which gets most of its revenue from network advertising.

In addition to spot advertising (where you purchase a set number of commercials) which are then played throughout a 24 hour period, talk radio stations also generate ad revenues from broadcasting syndicat-

In millions	1995	1996	1997	1998
Network	$480	$523	$559	$593
National Spot	$1,950	$2,135	$2,870	$3,129
Local Spot	$8,890	$9,610	$10,425	$11,315
TOTAL	$11,320	$12,268	$13,854	$15,037

Figure 10.7 Radio Advertising Revenues
Source: Television Bureau of Advertising

ed national shows. Examples of these programs include shows hosted by Howard Stern, John Boy and Billy, Rush Limbaugh and others.

Radio revenues represent only about 7% of all media generated revenue, including television, magazines, newspapers and yellow pages.

Revenues for radio advertising are growing approximately 5-7% annually with revenues expected to reach $15 billion by 1998. Figure 10.7 shows the growth in radio advertising revenues over the past 4 years.

Advertising rates vary widely for radio broadcast, depending on which type of advertising is chosen. These types include network, (where air times are chosen during specific shows, at a certain time, per commercial), spot (where run times are inserted during unsold network time slots) and local (where commercials run during unsold network time slots).

For local spot advertising, stations ordinarily sell packages which include a set number of commercials for a set price - 200 commercials for $1500 for example. This includes the production of the commercial by the station normally. For this, stations promise 2-3 spots will run during morning drive time, another 2-3 to run during the day, another 2-3 running during evening drive time and the balance to run from 7PM to 6AM. These spot ads act as filler ads to the station in an effort to fill in unsold time slots during the broadcasting day.

Supplementary Services

One of the most commonly used supplementary services for radio broadcast services is messaging. The best known form of messaging is paging service. FM subcarrier paging has been linked into a nationwide network in the United States and Canada by CUE Network Corporation and implemented on nearly 600 FM radio stations. CUE usually leases the rights to use the subcarrier from the radio station owner, installs a satellite dish to receive the nationwide paging messages and sells the paging service to customers, such as long distance truckers, who must have extensive coverage over wide areas. A monthly subscription for nationwide paging retails for $41.95, while Canada/US coverage is $49.95.

A supplementary service area that is growing in importance is datacasting. Datacasting is defined as "...the delivery of information from a central source to a large and potentially unlimited number of geographically-dispersed receiver sites at the same time [1]." This point-to-multipoint data distribution application highlights one of the strengths of subcarrier datacasting. The data content can be used to convey traditional information, such as news, traffic, weather and sports, or more esoteric information, such as gasoline prices for a group of service stations, or company intranet "push technology" narrowcasting to fill-in mobile workers web pages with dynamic updates. Airtime pricing for such data broadcasting is in the one cent per bit range for nationwide coverage, and around two tenths of a cent per bit for coverage in one major market, while retail pricing per customer typically varies widely according to the specific data services being used.

Future Enhancements

High Definition Television Digital Services

The specifications for HDTV digital systems allow for many types of data services in addition to digital video service. Digital HDTV channels carry high speed digital services that can be addressed to a specific customer or group of customers that are capable of decoding and using those services. Examples of these services include: special programming information, software delivery, video or audio delivery (like pay-per-view programming), and instructional materials.

The data rate available for additional services is dynamic and ranges from a few kbps to several Mbps, depending on the video and audio program content. The gross data rate of the HDTV system is 19 Mbps. The amount of this data rate that is used by the HDTV video signal depends on the compression technology. Video data compression produces a data rate that changes dependent on the original video signal. When the video program contains rapidly changing scenes, most of the

19 Mbps signal is required for transmission. If the video signal is not changing rapidly, much of the 19 Mbps can be used for other types of services.

Transmission of the additional services has a lower priority than transmission of the primary program. If the primary service (HDTV) consumes a large part of the data (such as a rapidly changing video action scene), the customer may have to wait for some time prior to receiving large blocks of data.

Digital Audio Radio Services (DARS)

Proponents of satellite-based digital audio radio services (DARS) scored a major victory when the Clinton administration succeeded in forcing the FCC to auction satellite DARS spectrum in 1997 as a means of raising up to $3 billion to help balance the budget. Results from the auction were disappointing, however, with bids totalling only $173 million. Indications are that digital broadcasting via satellite will commence before terrestrial digital broadcasts.

Chapter 11

Wireless Office and Cordless Telephones

Introduction

Traditional wired telephones such as office PBX systems and home telephones are converting to wireless devices. Wireless office systems are used to provide advanced telephone features within an office area. Cordless telephone systems provide basic telephone services within a residential area. Wireless office and cordless technologies have also begun to merge together with wide area wireless services such as cellular.

Wireless Office Telephone System (WOTS)

A wireless office telephone system (WOTS) is used in a business environment to provide features similar to a private branch exchange (PBX) for handheld or wireless desktop telephones. WOTS systems are also called wireless private branch exchange (WPBX) systems. They can be completely independent systems or they can be integrated with standard (wired) PBX systems. These systems have three basic parts: wireless telephones, radio base stations and a switching system.

WOTS are typically used to provide communication to office personnel and production workers who are highly mobile. These people are frequently away from their desk or other fixed telephone station locations. This group commonly involves such people as shop floor supervisors, technicians, retail salespeople and shipping managers. Wireless office systems permit workers to accomplish their mobile jobs and effectively communicate at the same time. Studies indicate that 10% to 15% of the personnel in a typical industrial or manufacturing location fall into this highly mobile category [1].

Wireless office telephones can be handheld devices or wireless desktop telephones. Handheld wireless office telephones are ordinarily smaller than cellular phones and have features similar to a wired PBX phone. Wireless office phones may have belt clips for easy wearing and may include vibrators to allow silent alerts while employees are in meetings. They often have antennas that are concealed within the plastic casing to allow a smaller size.

Wireless office base stations communicate with mobile and fixed wireless telephones that are operating in the system. Wireless office base stations are similar to cellular system base stations, as they are directly connected with the WOTS switching system by cable. This cable allows both communication data and power to be supplied to the base station. The primary differences between cellular base stations and wireless office base stations include lower power level and smaller size. Because the WOTS switching system has its own backup power supply, direct connection eliminates the need to have battery backup for each of the base stations.

The wireless office switching system coordinates the overall operation of the base stations and WOTS handsets. The wireless office switching system may be part of the wired PBX system (integrated) or may be a separate control or switching system (external). Integrated systems permit one switching system to serve all the needs of wired and wireless telephones connected to a company's phone system. An external system (stand alone) is used for an independent system or when a radio system is added as an extension to an existing system. When connected as an external system, the wireless office system is usually attached to an existing PBX through a proprietary connection line. Many manufacturers of PBX equipment use proprietary (unique) interconnection communication lines to allow for advanced services

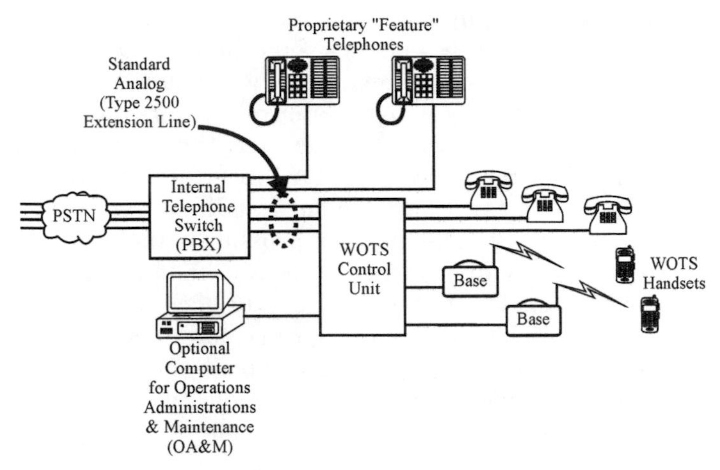

Figure 11.1 Wireless Office Telephone System (WOTS)

and in some cases to prevent other manufacturers from producing compatible equipment at lower cost. Interconnection lines allow for advanced features (such as message waiting indication and intercom caller identification) to be shared between the wired and wireless system. Wireless office switching systems normally include a control terminal for configuring and updating the WPBX with information about the wireless office telephones and how they can be connected to the PSTN.

Figure 11.1 shows a wireless office telephone system providing service to a number of wireless handsets and desktop units in a business environment. This wireless office system is connected to an existing system using a proprietary communications line. This wireless office system has the capability to provide service to both wired and wireless phone extensions.

Cordless Telephone Systems

Cordless systems are short range wireless telephone systems that are primarily used in residential applications. Cordless telephones regularly use radio transmitters that have a maximum power level below 10 milliWatts (0.01 Watts). This limits their usable range to 100 meters or less.

The earliest generation of home cordless telephones used a single radio channel that used amplitude modulation. These first generation cordless phones were susceptible to electrical noise (static) from various types of electronic equipment such as florescent lights. The noise encountered when using these phones sometimes created a consumer impression that cordless telephone quality was below standard wired telephone quality. This lead to the improved versions of cordless phones that used FM modulation to overcome the electrical noise. As cordless phones became more popular, interference from nearby phones became a problem. In apartment buildings where there were many users of cordless phones in close proximity, the ability to initiate and receive calls could be difficult as radio channels became busy with many users. This led to the development of cordless phones that used multiple radio channels. As voice privacy became more of an issue, cordless phones began to use scrambled voice. Some of these voice privacy systems were analog while a majority of cordless phones that offer voice privacy use digital transmission.

Until the mid 1990's, most cordless telephones were limited to use in a small radio coverage area of their base station that was usually located in the home. That home base station was normally connected to the telephone line of the owner (either residential or a single office telephone line) and they were not intended to serve the general public. To add more value to the use of cordless phones, cordless telephones evolved to allow access to base stations in public locations. Cordless telephones could then be used in the home and in areas which were served by public base stations.

The next evolution for cordless telephones was the combination of other types of wireless products and services into the cordless phone. This included the combination of wireless office and cellular telephones into a cordless phone. Figure 11.2 shows the evolution of cordless telephones.

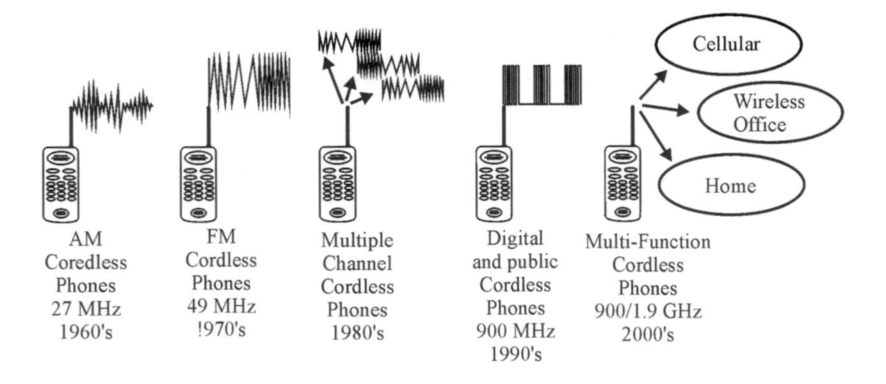

Figure 11.2 Evolution of Cordless Telephone Systems

Combined Cordless and Cellular

The most recent development of combined cordless and cellular systems will contribute to the rapid industry growth of the wireless office and cordless systems. The blending of office wireless systems (microcell systems) with the wide area cellular (macro cell systems) provides complete mobility services for enterprise accessibility, from the manufacturing floor, to the office, or on the road anywhere in the North American cellular network.

The early systems which most operated on the industrial scientific and medical (ISM) unlicensed 900 MHz frequency band were subject to interference, vast capacity constraints, and coverage limitations. Initially, only the Panasonic FreedomLink/ BusinessLink operated on cellular frequencies, offering the users the capability of using a single handset for both internal building and external wide area usage. The cellular carriers, however, assessed air-time or monthly usage fees on a per handset basis for "in-building" service, this coupled with cellular's unsophisticated "system" salesmanship contributed to limiting factor.

The situation and market has been evolving to where many of the systems are now operating in the cellular and (unlicensed PCS) 1920 - 1930 MHz bands, and well as offering tie-ins to the macro network. Many of the carriers, spurred on by new competition have relaxed their tariff structures for in-building usage, as well as have begun to realize the significance of in-building business users. Also macro wireless system and PBX manufacturers such as Nortel and Lucent are leading the integration of the private exchange (PBX) systems with the macro cellular networks. The results now permit an analog or digital cellular handset to be used both as a fully-featured business telephone within a facility and to take advantage of the macro cellular network when outside the facility. Thus a person can be reached on a single handset, anywhere in the North American cellular network.

The opportunities and applications presented for wireless office/ business telecommunications systems in the US marketplace are enormous. Improving communications is an essential component of a company's ability to booster productivity, meet customer needs, respond quickly to changing conditions and improve the quality of work life for employees. These days, few key employees remain at their desks to perform their duties. For such employees the time spent away from the office is often greater than the time spent in the office. Fixed telephones, even with call forwarding and voice mail features, are increasingly unsuitable to modern work styles, while portable phones are much more flexible and keep the communications with the person.

There are currently over 67 million business telephones in the US connected to PBX's, Centrex and Key Systems. Source: North American Telecommunications Association (NATA). A large portion of these wired phones will eventually migrate to wireless as markets develop and mature. The global market for In-building Wireless Business Telecommunications Systems by the year 2000 is estimated at $7 Billion.

Within the next two years the cost of a wireless Key System will approach that of a wired system. The average life of a Key System is seven years and that of a PBX is ten years. From 1998 onwards, it is expected that at least two percent of replacement Key Systems and one percent of PBX replacement systems will be wireless.

Market Growth

Both wireless office and cordless phones have experienced steady growth in units sold and dollar sales made. The technology of these systems is changing from analog to digital radio transmission. The improved quality of digital wireless and advanced features are projected to continue the growth trend for new systems as well as a very strong equipment replacement market.

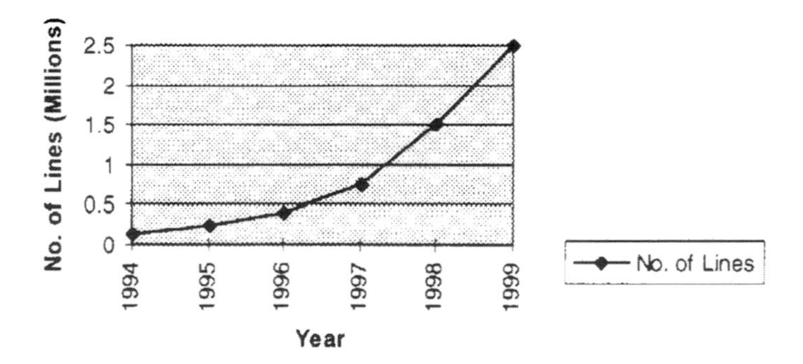

Figure 11.3 Wireless Office Market Growth
Source: EMCI (now Strategis Group)

A majority of office telephone systems continue to use wired systems. However, the growth of the wireless office systems has been above 40% per year and market projections show continued strong growth. The key factors for market growth increases include the conversion to private digital wireless systems, new multi-mode (office, cellular and cordless) and low cost end user equipment. Figure 11.3 shows the market growth for wireless office systems in the United States..

Sales of cordless phones in the United States increased from 19.5 million units sold in 1995 to 20.3 million units sold in 1996. Total dollar sales in 1996 was approximately 1.1 billion dollars. Sales are projected to increased over the next 4 years as a result of technological advances made in 1996 which include increased spectrum allowance, longer range cordless phones, digital transmission and advanced features. Figure 11.4 shows the increased expansion in sales and number of units for the past 10 years for cordless phones.

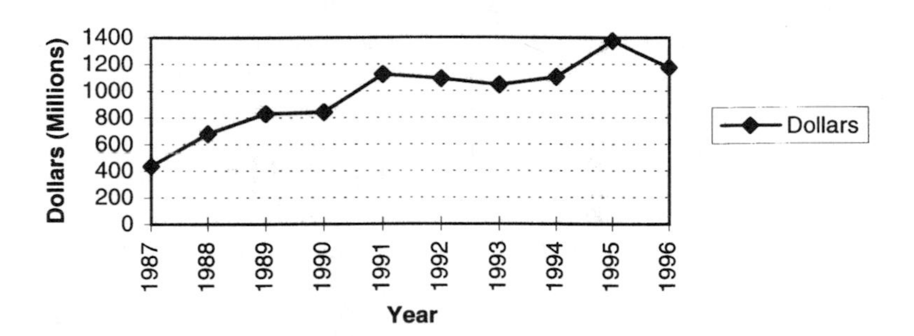

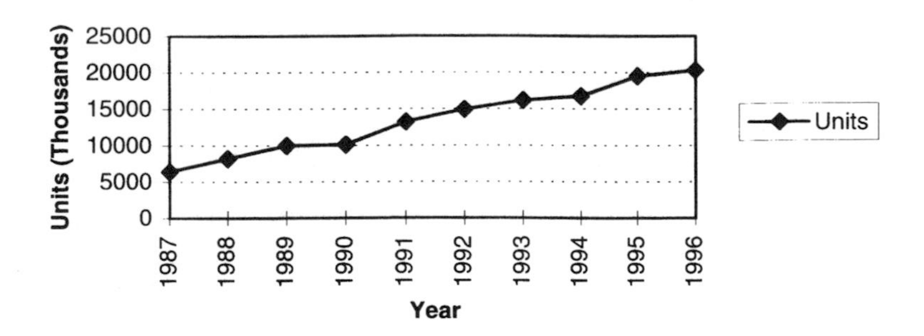

Figure 11.4 Cordless Market Growth
Source: EIA

Technologies

The key technologies used in wireless office and cordless telephone systems include independent interference avoidance, multi-mode capability and digital service.

Independent Radio Channel Coordination

Residential cordless telephones must automatically coordinate their radio channel access as they operate independently of any type of network control. To coordinate radio channel access and avoid interference to other cordless handsets installed in the vicinity, cordless phones perform radio channel scanning and interference detecting prior to transmitting a signal.

Because cordless telephone systems do not as a rule have a dedicated control channel to provide information, the cordless handset and base station continuously scan all of the available channels (typically 10 to 25 channels). Figure 11.5 (a) shows this process. When the cordless phone or base station want to transmit, the unit will choose an unused radio channel and begin to transmit a pilot tone or digital code with a unique identification code to indicate a request for service (figure 11.5 (b)). The other cordless device (base station or cordless phone) will detect this request for service when it is scanning and its receiver will stop scanning and transmit an acknowledgement to the request for service. After both devices have communicated, conversation can begin (figure 11.5 (c)). When another nearby base station detects the request for service, it will determine that the message is not intended for it and will not process the call and scanning will continue.

When there are many cordless telephones that attempt access in the same area, the problem of excessive radio channel interference will result in two likely outcomes; the automatic inhibiting of transmission (phone displays busy or provides a busy tone) or the phone will operate but there will be an unacceptable level of interference. For analog handsets, interference may be cross-talk (conversations from other phones) and for digital handsets, there will be distorted audio.

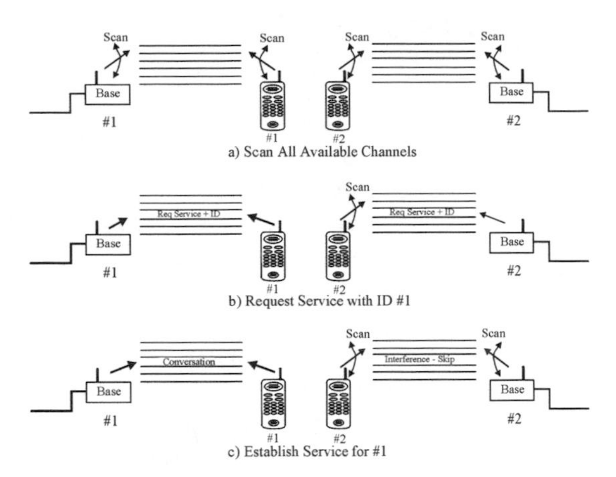

Figure 11.5 Independent Radio Channel Coordination

Multi-Mode

Multi-mode telephones have a combination of multiple wireless technologies in a single wireless handset. This may include the combination of wireless office, cordless and wide area cellular types of services, in addition to analog or digital transmission capability. Because many of the new wireless technologies involve digital transmission, the combination of multiple technologies into one handset is simplified. Much of the digital signal processing can be performed by the same microprocessors or digital signal processors (DSPs) in the handset.

There are some key challenges with multi-mode telephone service. These include differences in feature operation and performance changes between different systems. The operation of a wireless telephone in a wide area network such as cellular is different than a operation in a wireless office or cordless environment. For example, residential systems do not provide a dialtone, wireless office systems often require the dialing of a prefix (e.g. 9) to get a dialtone, and cellular systems do not provide any dialtone. Multi-mode telephones would operate differently depending on which system they are operating in.

There will also be some performance differences between systems. These include shorter battery life when the wireless telephone is used in the cellular system compared to when it is used in a low power cordless environment.

Digital Transmission

Many cordless phones have started to use digital radio transmission to allow for consistent voice quality, voice privacy and to provide for advanced services. Digital transmission offers the ability to correct for some of the errors that result from radio transmission and provides a better way to detect if interference is occurring. This allows for more consistent voice quality. Digital signals can be easily manipulated with secret codes that allow for increased voice privacy. Digital transmission also allows for flexibility of services. For example, if a new feature such as voice mail message waiting notification needs to be added, this may only require a software change in the handset to decode the message when it is received. The decoding of the message may simply place a digit on the screen of the wireless telephone indicating how many messages are waiting.

Commercial Systems

There are many commercial cordless and wireless office systems. Some of the more popular systems include Tele-GoTM, CT-2, PHS, IS-94, SpectraLink, Companion and DECT.

Tele-GoTM

TeleGoTM is a dual-mode cellular and cordless analog system that operates in the 800 MHz cellular frequency band. TeleGoTM telephones can communicate with either a cellular system or a home cordless base station. When the TeleGoTM telephone is operating within the radio coverage area of the home base station, the user can make and receive calls via a standard telephone line (local telephone system). When the user carries the handset outside of the range of its cordless base, the user can make and originate calls via the public cellular system.

The TeleGoTM system consists of a home cordless base station, multimode handset and a home cellular system that has software to allow the sending and automatic forwarding of calls to and from the cordless base station. The cellular system call processing software allows the cellular switching system to communicate with the TeleGoTM cordless base stations.

The TeleGoTM system uses standard 30 kHz AMPS radio channels and 10 kHz NAMPS radio channels to communicate between the home cordless base station and TeleGoTM handset. The home cordless base station can be programmed by messages that are sent from the cellular system. This allows the selection of its operating frequency and it also allows the cellular system to disable the operation of the home cordless base station in the event the customer does not pay their telephone bill.

The TeleGoTM user is assigned a single telephone (directory) number which is the phone number assigned to the cellular system. The cordless base station has a different telephone number. When the cellular system detects that the TeleGoTM telephone is operating at home (the cordless base calls the cellular system after it has detected the presence of the TeleGoTM handset), it automatically forwards the calls to the TeleGoTM home base station. When the TeleGoTM handset goes out of range of the home base station, the call forwarding feature is stopped and calls are routed through the cellular system.

Several TeleGoTM handsets can be configured to work with different cordless base stations. This allows several members of a group of office workers to each have their own TeleGoTM handset.

Cordless Telephony 2nd Generation

CT-2 is a wireless telephone system that allows cordless telephones to be used at home and in public places such as shopping malls, hotels, and train stations. The popularity of home cordless telephones and a lack of public coin telephones in the United Kingdom, during the 1980s, accelerated the development of the public cordless telephone service. The concept of a public cordless telephone service was officially designated CT-2.

In 1988, three CT-2 competitive firms began to offer service in the UK. These companies did not initially offer compatible systems. Eventually, these firms recognized the importance of radio coverage and the key disadvantage that incompatibility between systems significantly reduced the benefit of public cordless service. This led to the development of the single cordless telephone system specification that permits the use of different phones in different systems. Because of delays in standardization, all three original British competitors are now no longer in the CT-2 service business. However, CT-2 equipment is in use in other countries such as Hong Kong and Canada.

The basic parts of the CT-2 system are a dual function wireless telephone which can operate with a home base station or public cordless base station. The handset is called the cordless portable part (CPP) and the base station is called the cordless fixed part (CFP). The CFP can either be a home fixed part or a public fixed part. The home CFP allows the handset (CPP) to originate and answer calls. The public CFP only allows the handset to originate calls. The CT-2 evolved to the CT-2+ system which also allowed the public CFP to originate and answer calls.

CT-2 systems use extremely low radio transmit power (maximum 10 milliwatts). Because of this low power, it is only possible for the CT-2 system to operate at distances of up to a few hundred meters from the base station. The RF power may be reduced by 16 dB by a command that is received from the base station. This allows the CT-2 to minimize interference to other handsets in a crowded urban environment.

The CT-2 systems uses 32 kb/s ADPCM speech coder that provides audio quality which is similar to the quality received via standard wired telephones. Unfortunately, the design of the digital transmission system was not optimized for use in rapidly moving vehicles (such as cars and trains). Because there is no digital error protection, the use of CT-2 phones on rapidly moving vehicles commonly results in poor (distorted) quality audio signal.

The CT-2 system uses time division duplex (TDD) operation. In the TDD system, the base station transmits for approximately 1 msec and the cordless handset replies with a 1 msec burst. Each 100 kHz single CT-2 radio channel can service one conversation at a time. Each 100 kHz radio channel has a combined data transfer rate 72 kb/s. Because of the TDD operation, this provides 36 kb/s data transfer in each direction. The modulation method is frequency shift keying (FSK) of the GMSK type.

The original CT-2 systems did not allow for handoff between radio coverage areas. This required the user to remain in the radio coverage area for the entire duration of the call. This is ordinarily adequate for home use; however, it is undesirable for larger high traffic public areas such as airports, train stations, or shopping malls.

The standard frequency band used for CT-2 radio service is 864 MHz to 868 MHz. This provides for up to 40 radio channels. The CT-2 standard has evolved into several other industry standards which operate on different frequencies. These frequencies include 900 MHz unlicensed frequency band and the 1900 MHz PCS band.

Personal Handiphone System (PHS)

Personal Handiphone System (PHS) is a digital cordless telephone system that allows wireless telephones to be used at home and in public places. Since its introduction in Japan, the number of PHS customers has grown to over 4 million in 1997 with more than 300,000 microcell base stations. In 1998, PHS systems were operating or testing in over ten Asian countries.

Similar to other cordless technologies, PHS handsets are small and have long battery life; over five hours of talk time and over 550 hours of standby time. Some PHS phones have built in cameras. PHS videophones have been demonstrated and voice activated wristwatch phones were in use in 1997 [2].

PHS handsets have a maximum transmit power level of 10 milliWatts. This allows the handset to operate only within a few hundred meters of the base station.

The PHS systems uses a high quality 32 kb/s ADPCM speech coder that provides audio quality comparable to standard wired telephones. As with the CT-2 system, the design of the PHS digital transmission system was not optimized for use in rapidly moving vehicles. This limits the use of PHS telephones to slow moving vehicles or walking pedestrians.

PHS radio channels use TDMA/TDD access technology with four communication channels (time slots) per radio channel. The designated frequency bands in Japan are 1895 MHz to 1906 MHz for indoor and 1895 to 1918 MHz for outdoor coverage areas. This allows a total of 77 radio channels for the frequency bands. Different frequency bands may by used in other countries [3].

IS-94

The interim standard 94 (IS-94) system is a wireless office telephone system that uses analog (AMPS) cellular technology. There are several different brand names used by manufacturers that market IS-94 compatible systems. These include OffiCell (office version) and ContiNet (home version) by Astronet (Mitsubishi), BusinessLink by Matsushita (Panasonic). A similar competing system is the interim standard 91A (IS-91A)

IS-94 systems use private setup channels (control channels) when in use in a business office which permits the private base station to command wireless office telephones to use private voice channels. These private setup channels are not one of the standard control channels reserved for the public cellular network. Instead, some radio channel frequencies are used by the Business Link system which are not used by the public cellular network in that area. Similarly, the voice channels used by the private system are carrier frequencies not used in that cell or area by the public cellular network.

An important feature of all of these systems is the ability of the handset to operate on both the private and the public cellular systems. Private systems use very low RF transmit power levels so they do not usually cause interference with the public cellular network.

An IS-94 system is a composite of cellular and wireless office systems. Each of these independent systems can work with the same single wireless handset. This allows IS-94 handsets to communicate with either cellular systems, office systems or special home base stations. IS-94 phones can originate and answer calls on the public and private networks.

IS-94 can use the narrow band NAMPS 10 kHz analog FM mode for voice channel transmission to give more radio channel capacity. The output power of IS-94 handsets is very low when operating with office or home base stations. This limits the radio coverage area from each base stations to less than 500 meters in diameter. The IS-94 system does allow handoff in the wireless office environment which permits very large radio coverage areas in an office or industrial environment.

IS-94 private systems share radio channels with the cellular systems. This requires the IS-94 system to check the use of radio channels prior to transmitting on a voice channel. IS-94 sets are pre-programmed with a list of private radio channels to help them find the private setup channels. The automatic scanning capabilities of the IS-94 system will prevent that system from using a radio channel when there is already an RF signal there from the public cellular system. The public cellular system operator may reserve certain channels for a particular IS-94 system operation to ensure minimal interference. IS-94 systems can be adjusted by remote (dial in modem) to make changes to frequency assignments and key features.

The IS-94 system uses the AMPS frequency bands; 824 to 849 MHz for the reverse link (telephone to base) and 869 to 894 MHz for the forward link (base to telephone) with the exclusion of the standard AMPS control channel frequencies. IS-94 base stations can use 30 kHz wide or 10 kHz narrowband AMPS (NAMPS) channels.

SpectraLink

SpectraLink is a company that manufactures wireless office telephone systems. SpectraLink developed the radio interface and necessary PBX system interfaces to allow wireless office telephones to emulate the operations of the popular PBX systems.

SpectraLink equipment uses a lower power TDMA Frequency Hopping (FH) radio interface. The system uses a combination of TDMA and frequency hopping (jumping to a different carrier frequency for each time slot) to allow operation of multiple handsets in an office area with minimal interference. The SpectraLink system selects the sequence of frequencies used for FH which appears random (called "pseudo-random"). But the sequence is really known to both the base radio and the handset so the two can remain in communication. While the hopping sequence from nearby wireless handsets may lead to an occasional "collision" between two different handsets, the brief interference on a single time slot does not decrease audio quality significantly.

SpectraLink systems primarily use unlicensed frequency bands. Other radio systems may also be operating in the bands. By properly programming the mobile units when they are installed in the system, the selection of frequency hopping frequencies can be adjusted to minimize or ensure local sources of interference (for example, from a wireless LAN device) do not operate in the same band.

Companion™

Companion™ is an upgraded CT-2 cordless system developed by Nortel. The first Companion product was introduced in Hong Kong in 1992. In 1996, a multi-mode Companion Microcellular system was introduced which combined a cellular phone with Companion technology. By allowing a Companion handset to appear as a PBX extension, this allows calls to be simultaneously routed to the wireless handset and the wired PBX telephone extension. It is also possible to use a Companion handset with a cordless base station in the home.

The Companion system is usually described as CT-2+ because of the improvements over the original CT-2 system. The most significant improvement was the ability to provide handoff between base stations.

The Companion consists of wireless handsets, base stations and a controller. The controller can normally be connected to other wired PBX and key telephone systems.

Companion handsets can operate on CT-2 systems. The Companion system operates in a variety of radio frequencies depending upon the country. The standard frequency band used for CT-2 radio service is 864 MHz to 868 MHz. In Canada, the Companion system operates on the 948-952 MHz and in the US, Companion operates on the 1. 9 GHz unlicensed PCS frequency band.

Digital European Cordless Telephone (DECT)

The DECT system is a digital wireless office telephone system. DECT was originally developed by the European Telecommunications Standards Institute (ETSI) technical standards committee in the late 1980s and the specification was released in 1992. It was first intended that the use of the DECT system was for wireless office. After its release, it was adapted to allow low-tier PCS and home cordless use as well.

There are two version of DECT: the European version and the American version. The European version uses a very wide radio channel to allow up to 12 simultaneous wireless telephones to share each channel. The American version uses a slightly more narrow radio channel and allows up to eight users to share a single radio channel. Personal Wireless Telecommunication (PWT) is an adaptation of DECT for the North American market.

The DECT system has been adapted to several different frequencies and data rates and is also being used in the US PCS frequency bands.

The DECT system consists of wireless handsets, data terminals, base stations, a wireless switching system (PBX), an interface to the wired telephone network and sometimes a link to a computer/data network.

DECT is easily adaptable to provide data communications such as a wireless local area network.

The DECT system uses TDD multiplexing which allows the same radio channel frequency to transmit and receive. The transmitter and receiver bursts are grouped together into frames. A single time slot can provide 32 kb/s data for each handset. Multiple time slots can be grouped together to supply much higher data rates for data products such as a computer network devices. While the DECT switching system controls the overall operation of the base stations, DECT base stations and handsets independently control their use of radio frequencies. DECT base stations can automatically change their frequencies to avoid interference.

DECT base stations normally have two antennas for diversity reception. One of the key advantages of the DECT system is its ability to perform a seamless handoff. Two base stations can simultaneously communicate with the DECT handset at the same time. This process involves using different time slots on each of the base stations to communicate with the DECT handset during handoff.

In Europe, DECT carrier frequencies are reserved on the 1880-1900 GHz band, with 10 carriers starting at approximately 1882 MHz and ending at 1897 MHz, spaced 1728 kHz apart. Because DECT and PWT are TDD systems, the same frequency is used for the uplink and downlink. In North America, the bandwidth of PWT is only 864 kHz due to the use of the more radio spectrum efficient DQPSK modulation. The PWT carrier frequencies are usually located in the U-PCS band starting at 1920 MHz and ending at 1930 MHz.

The maximum RF Power of PWT handsets is 200 milliwatts. They are commonly commanded by the base station to use a lower operational power level, down to as low as 5 milliwatts. Base station power can be one watt or more to increase cell coverage. However, the general objective in a DECT system is to use small cells of less than 1 km (0. 6 mi) diameter.

Services

Most wireless office and cordless systems provide for voice, data and advanced PBX features.

Voice

Cordless and wireless office systems ordinarily offer high quality voice services in a slow moving environment. Some of these systems have combined with cellular service to allow wide area coverage. Because cordless and wireless office systems normally use unlicensed frequency bands, the customer has unlimited use of the radio spectrum without charge. In some shared systems that use licensed frequencies, the service provider may charge a monthly access fee and provide unlimited usage when the phone is operating in the wireless office or home environment.

Data

The basic data rates for wireless office and home cordless systems ranges from 32 kbps to over 700 kbps. This allows some wireless office systems to also serve as a wireless local area network (WLAN).

PBX Functionality

Wireless office systems regularly have the capability for advanced PBX features such as three or four digit dialing, voice mail notification and different ringing for intersystem calls.

Public Area Mobility

The trend in many wireless office and cordless products is to combine them with wide area systems such as cellular and PCS.

Future Enhancements

The enhancements to wireless office and cordless systems include advanced features with higher data rates. These types of services may include Internet access or video display service.

Internet Access

A wireless access protocol (WAP) forum was established in 1997 to standardize how Internet content and advanced services can be utilized by digital wireless phones and other wireless devices. The goals of the WAP forum include the creation of a global wireless protocol specification that allows applications to operate on wireless devices independent of their type of network access technology (GSM, CDMA, TDMA, DECT, PHS). The first WAP specifications became available to the public in February 1998 and are available at www.WAPforum.org.

The key limitation of wireless information services is the limited amount of data transmission capability offered by mobile systems. WAP defines Internet and other information application services that have more efficient, simplified or compressed communication requirements. WAP has defined: a Micro-Browser, similar to other Internet browsers; language scripting similar to JavaScript to provide means for dynamically enhancing mobile device capabilities and allow access to telephone services. WAP also reviews content formats such as calendars (vCalendar) and business cards (vCard).

1. "Cellular and PCS, The Big Picture," Harte L., Levine R., McGraw-Hill, 1997.
2. "The rise and rise of Japan's PHS," Mobile Communications International, IBC Publishing, London, United Kingdom September 1996, pg 47.

INDEX

Fax	1-800-825-7091 1-919-557-2261	Internet	Success@APDG-Inc.com www.apdg-inc.com
Telephone	1-800-227-9681 1-919-557-2260	Mail	202 N. Main St. Fuquay-Varina, NC 27526

Please Send The Following Books:
I understand that I may return any book for a full refund within 30 days of purchase.

Book #	Title	Price
	Subtotal	
	Shipping and Tax	
	Total	

Sales Tax:
North Carolina customers please add 6% sales tax

Company:______________________________

Name:______________________________

Address:______________________________

City:______________ State: ______ Zip:_______

Telephone:______________ Fax:______________

Shipping and Handling:
$5.00 per book in the U.S. and Canada
$15.00 per book outside the U.S. and Canada

Payment:

☐ Check ☐ Visa ☐ MC ☐ Amex ☐ Diner's Club

Card Number:______________________________

Name on Card:________________ Exp. Date:_______

☐ **Please send me a free catalog**

Fax	1-800-390-5507 1-919-557-2261	Internet	Success@APDG-Inc.com www.apdg-inc.com
Telephone	1-800-227-9681 1-919-557-2260	Mail	4736 Shady Greens Drive Fuquay, NC 27526

Please Send The Following Books:
I understand that I may return any book for a full refund within 30 days of purchase.

Book #	Title	Price
	Subtotal	
	Shipping and Tax	
	Total	

Sales Tax:
North Carolina customers please add 6% sales tax

Company:____

Name:____

Address:____

City:____________________ State: _______ Zip:________

Telephone:___________________ Fax:___________________

Shipping and Handling:
$5.00 per book in the U.S. and Canada
$10.00 per book outside the U.S. and Canada

Payment:

☐ Check ☐ Visa ☐ MC ☐ Amex ☐ Diner's Club ☐ Discover

Card Number:__________________________________

Name on Card:____________________ Exp. Date:________

☐ **Please send me a free catalog**